# Mind Math

## Math Activities Help Kids Sharpen Their Mind

*(A Quick and Easy Guide to Mental Math and Faster Calculation)*

**Joan Vargas**

Published By **Darby Connor**

# Joan Vargas



*Mind Math: Math Activities Help Kids Sharpen Their Mind (A Quick and Easy Guide to Mental Math and Faster Calculation)*

ISBN  978-1-77485-684-0

No part of this guidebook shall be reproduced in any form without permission in writing from the publisher except in the case of brief quotations embodied in critical articles or reviews.

Legal & Disclaimer

The information contained in this ebook is not designed to replace or take the place of any form of medicine or professional medical advice. The information in this ebook has been provided for educational & entertainment purposes only.

The information contained in this book has been compiled from sources deemed reliable, and it is accurate to the best of the Author's knowledge; however, the Author cannot guarantee its accuracy and validity and cannot be held liable for any errors or omissions. Changes are periodically made to this book. You must consult your doctor or get professional medical advice before using any of the suggested remedies, techniques, or information in this book.

Upon using the information contained in this book, you agree to hold harmless the Author from and against any damages, costs, and expenses, including any legal fees potentially resulting from the application of any of the information provided by this guide. This disclaimer applies to any damages or injury caused by the use and application, whether directly or indirectly, of any advice or information presented, whether for breach of contract, tort, negligence, personal injury, criminal intent, or under any other cause of action.

You agree to accept all risks of using the information presented inside this book. You need to consult a professional medical practitioner in order to ensure you are both able and healthy enough to participate in this program.

# TABLE OF CONTENTS

# Introduction

I cannot describe in words how excited I was the day I first saw how to multiply any two-digit number by the number 11 faster than a calculator. Let me show you. Let's take for instance I wish to multiply 53 by 11. It is as easy as adding the digits of the two-digit number together and dropping it in their middle. So 5 + 3 = 8, and the answer to 53 × 11 has to be 583. Impressive, isn't it?

This little trick or call it shortcut was what turned me to a math nut and as James Randi said, 'Immunization against such afflictions are not available. You have to recover all by yourself. Beware.' After learning the trick, the next thing I learned was how to square any number that ends in 5 faster and some other mind blowing magical feats.

My curiosity to know more soon overcame me and it became so unbearable until I delve into the world of studying numbers and the beautiful pattern they are capable of displaying. In the middle of my ardent quest to find out better and faster way of computing and calculating numbers, I came across one of the most important person in my life. He is my distant teacher and a professor of mathematics at Harvey Mudd College, Claremont, California. He is my greatest discovery outside of my relationship with God. He is the brain behind most of the procedures and processes laid out in this book and most mathemagic tricks of today. He popularized the term "mathemagician". He is a mathematics teacher and of course a mathematics entertainer better known as a mathemagician. He is Professor Arthur Benjamin. I read two of his books – 'secrets of mental math' and 'the magic of math'. I also watched a lot of his videos and from there I learnt most of the shortcuts for doing mental math as presented to you in this book. This is why I don't see myself as a genius or a prodigy because what I can do, I learnt from someone, and if anyone also cares enough to learn the way it is done and put in a lot of effort and practice, he would do exactly what I'm doing with relatively the same speed or even faster.

This book will show you how to round a few of the corners and cut through some of the traffic while dealing with arithmetic. One of the things that dawned on me while studying most of the mental math shortcuts was as I worked through the shortcuts I could see other ways of calculating faster and easier and with just a little bit of effort I could discover or even create new methods of my own which adds more excitements to my journey through this corner of the world of numbers.

The book is titled: Mentally calculate it – gateways to becoming a human calculator. But I must warn that the shortcuts for doing mental math provided in this book are not primarily intended to replace the electronic calculator. I strongly believe there is a place for both. For long and complicated computations, the calculator has no equal, but for some simpler problems, a rapid mental calculation can often give the answer correctly faster than you can enter the function on a calculator. The aim of this book therefore is to lower your dependence on calculator and enhance your self-confidence at using your God-given devices (the mind and the brain) to compute simpler problems. It is therefore my deepest hope that this book will assist towards a better understanding of the basics involved and a deeper appreciation of the wonders of playing with and manipulating numbers.

# Chapter 1: Mental Additions And Subtractions

LEFT – TO – RIGHT ADDITION

We pronounce numbers form left – to – right, we write numbers from left – to – right and wouldn't it be natural if we could learn to calculate numbers from left – to – right instead of the right – to – left method which we were taught in school. When you use the left – to – right method for arithmetical calculations, you'll find out that you can call out your answer before your classmates can put down their pencils on paper. And you don't even need a pencil!

In this chapter you'll learn the left – to – right method of doing mental addition and subtraction for most numbers ranging from two – digit to three – digit; to even four – digit numbers and above with lightning speed.

Why use the left – to – right method for Arithmetical Operations?

1.The first reason is that to do mental math, the best method suitable for doing it is the left – to – right way. The right – to – left method may be good or even best on paper but if you really want to do mental math you have to master your calculations from left – to – right.

2.The second reason is that doing calculations from left – to – right gives you a possible and realistic estimate of the answer to your problem. What I mean is, it's best to know that your answer is "a little over 7500" than to know that it "ends in 6". Thus, by working from left to right you begin with the most significant digits in your problem.

3.The third reason for doing calculations from left – to – right (which is why I love it) is because it gives you the opportunity of saying your answer while you're still calculating in your head.It makes you look like a genius without really trying. To most people it will be like you instantaneously blotted out the answer as soon as you were given the problem.

Given these three reasons, the earlier you get used to computing from left – to right, the better.


TWO DIGIT ADDITIONS


The assumption here is that you already know how to add and subtract one digit numbers. We will begin with two – digit additions. Also in this chapter you'll learn how to "simplify" your problems by breaking it into smaller, more manageable parts. Also two digit addition can be so easy when it doesn't involve a carry.
 For example:


To solve 36 + 23, first add 20, then add 3, after adding 20, you have the problem 56 +3, which equals to 59. It can be illustrated as follows;
 36 + 23   =56+ 3= 59


The above diagram is just a way of representing the mental process involved in arriving at the answer using the method. Let's try another example, this time the one which requires you to carry;

Adding from left to right, you can simplify the problem by adding 48 + 20 = 68,          then 68+ 9 = 77.

  48 + 29 = 68 + 9    = 77


Now try this one on your own, mentally calculating from left to – right, and then check under it to see how it is done.


  97 + 36 = 127+ 6= 133


If you have problem with carrying numbers, don't worry about it. You'll soon get used to it and I can assure you that with just a little bit of practice you'll get better at doing it. Now, try another problem for practice, again computing it in your mind first, and then check how it is done.

You should have added 68 + 40 = 108, and then 108 + 5 = 113, the final answer. Was that easier? If you would like to try your hand at more two – digit addition problems, check out the set of exercises below.

Exercise: Two – Digit Additions

1. 3. 5.


2. 4. 6.


THREE DIGIT ADDITIONS


The strategy for computation (from left – to – right) remains unchanged. After each step here, you'll arrive at a new and simpler addition problem. You'll love it.


 Example:

Starting with 639, we add 200, then add 30, then add 7. After adding 200 (639 + 200 =839), the problem becomes 839 + 37. After adding 30 (839 + 30 = 869), the problem simplifies to 869 + 7 = 876. This thought process can be diagrammed as follows;

 639 + 237 = 839 + 37=869 + 7  = 876


All mental addition can be done using this method. The goal is to keep "simplifying" the problem until you are just adding a one – digit number. Try this other addition problem in your mind before looking at the working process that follows;

Did you reduce and simplify the problem by adding left to right? After adding the hundreds (447 + 100 = 547), you were left with 547 + 69. Next you should have added the tens (547 + 60 = 607), simplifying the problem to 607 + 9, which you then summed to get 616.
Diagrammatically, the problem would look like this;

447 + 169 = 547+ 69 = 607 + 9= 616

Note: When you are doing the problems for the first time, practice them out loud. Reinforcing verbally will help you learn the mental method much more quickly.

Now, a harder problem,

Now, look to see how we did it.

858 + 634= 1458  + 34 =1488 + 4   = 1492

Let's try one more for practice:

Do it in your mind first, and then check the computation below;

659 + 596= 1159 + 96 = 1249 + 6= 1255

This addition problem is a little more difficult than the previous ones since it requires you to carry numbers in all three steps. However, for these kinds of problem, we have an alternative and simpler method. You will agree with me that it is a lot easier to add 600 to 659 than it is to add 596, so try adding 600 and then subtracting the difference.

659 + 596= 1259 − 4 = 1255

And even sometimes you'll need to switch numbers. Look at this next problem:

307 + 328= 328 + 307=628 + 7=635

I can guarantee you that with just a little bit of practice, you'll get really good at doing additions mentally, not just for two − digit or three − digit numbers but also for larger digits.

Note: When doing mental additions:

☐      Look for opportunities to combine numbers to reduce the number of steps to solution (check examples).

☐      Look for opportunities to form 10, 100, 1000 and so on between numbers that are not necessarily next to each other (check examples).

☐      When doing a problem, it is necessary to think only of the results of each step and omit any unnecessary details of the step itself. For instance, in adding 6 + 13 + 7, you'll slow up yourself by stopping to say "6 plus 13 equals 19, plus 7 equals 26." Instead, just say: "19, 26." Omit the words of a process as much as possible and concentrate on the results.

☐      Look for opportunities to switch between numbers by placing the easier one next to the harder one; this was done in the last problem to ease the mental effort.

Final Note:PRACTICE!

EXERCISE:THREE DIGIT ADDITION

1. 3. 5.

2. 4. 6.

## LEFT – TO – RIGHT SUBTRACTIONS

Talking about subtraction, most people find it a lot easier to add than to subtract but if you can do your subtraction as well from left to right, it'll become almost as natural and easy as addition.

## TWO – DIGIT SUBTRACTIONS

When subtracting two – digit numbers, your goal is to simplify the problem so that you are reduced to subtracting (or adding) a one – digit number. Let's begin with a very simple subtraction problem.

After each step, you arrive at a new and easier subtraction problem. Here, we first subtract 20 (96 – 20 = 76), then we subtract 5 to reach the simpler subtraction problem 76 – 5 for the final answer of 71.

$$96 - 25 = 76 - 5 = 71$$

Of course, subtraction problems can be a lot easier when there is no borrowing (which occurs when a larger digit is being subtracted from a smaller one). But the good news is that "seemingly hard" subtraction problem can usually be turned into "easy" addition problems. For example:

These are two different ways to solve this problem mentally.
1. First subtract 20, then subtract 9
$$86 - 29 = 66 - 9 = 57$$

But for the problem, I would prefer the following strategy. I call it the "over – subtraction method"!
2. First subtract 30, and then add back 1
$$86 - 29 = 56 + 1 = 57$$

Don't panic! Here is a simple rule guiding the choice of which method to use. If a two – digit subtraction problem would require borrowing, round the second number up (to a multiple of ten). Subtract the rounded number, and then add the difference. For example, 74 – 38 would require borrowing (since 8 is greater than 4), so round 38 up to 40, compute 74 − 40 = 34, then add back 2 to get 36 as your final answer.

74 − 38 =  34 + 2 = 36

Now try your hand (or head) at 91 − 47. Since 7 is greater than 1, we round 47 up to 50, subtract if from 91 (91 − 50 = 41), then add back the difference of 3 to get the answer.

91 − 47   = 41 + 3=44

With just a lIttle bit of practice you will become comfortable working subtraction problems both ways. Just use the rules above to decide which method will work best.

EXERCISE:TWO DIGIT SUBTRACTION
1. 3. 5.

2. 4. 6.

THREE DIGIT SUBTRACTIONS

If you have worked and mastered the two – digit subtraction using both ways illustrated above, you can then proceed to three – digit subtraction problems.

This problem doesn't actually require you to borrow any numbers (since every digit of the number is less than the digit above it). Simply subtract one digit at a time, simplifying as you go.
 $958 - 417 = 558 - 17 = 548 - 7 = 541$

Now let's try another three digit subtraction problem that requires you to borrow a number.

At first glance, this problem look like a pretty tough problem, but if you first subtract $847 - 700 = 147$, then add back 2, you reach your final answer of $147 + 2 = 149$.
 $847 - 698 = 147 + 2 = 149$

Now try one yourself:

Did you first subtract 500 from 953? If so, did you get $953 - 500 = 453$? Since you subtracted 8 too much, did you add back 8 to reach 461, the final answer?

953 − 492 =453 + 8 = 461


Now, if you check all these problems you'll find out that they include numbers that were close to a multiple of 100 (Did you notice?) But what about problem like:


If you subtract one digit at a time, simplifying as you go, your sequence will look like this;
875 − 458 =                475 − 58        = 425
− 8          = 417


But what happens if you round up to 500?
875 − 458=375 + ??=??


Subtracting 500 is easy: 875 − 500 = 375. But you have subtracted too much. So you'll need to figure out exactly how much too much. At first glance, the answer seems to be far from obvious. To find it you need to know how far 458 is from 500. The answer can be found by using "complements", a very effective technique that will make three – digit subtraction problems a lot easier to do.

COMPLEMENTS

Quick, how far from 100 are each of these number?

5846347679

Here are the answers:

 100 100  100 100 100

Note that for each pair of numbers that add to 100, the first digits (on the left) add to 9 and the last digits (on the right) add to 10. We then say that 42 is the complement of 58, 54 is the complement of 46, and so on.

Now, you try to find the complement of these two – digit numbers.

3759934408

To find the complement of 37, first figure out what you need to add to 3 in order to get 9. (The answer is 6). Then figure out what you need to add to 7 to get 10 (The answer is 3) Hence 63 is the complement of 37.

The other complements are 41, 07, 56, and 92. Notice that, like everything else you do as a human calculator, the complement must be gotten from left to – right. As we have seen, the first digits add to 9, and the second digits add to 10. (But an exception to this occurs in numbers ending in 0e.g. 30 + 70 = 100 – but these complement are way too easy!)

So, what do complements have to do with mental subtraction? Well, I'm glad you asked. They allow you to convert difficult subtraction problems into straight forward addition problems. Let's revisit the last subtraction problem that made us perplexed.

To begin you subtract 500 instead of 458 to arrive at 375 (875 – 500 = 375). But then having subtracted too much, you needed to figure out how much to add back. Then here comes the use of complements to save the situation. Using complements gives you the answer in a flash. How far is 458 from 500? The same distance as 58 from 100. If you find the complement of 58 the way we have shown you, you'll arrive at 42. Add 42 to 375 to arrive at 417, your final answer.

875 – 458 =375 + 42 = 417

Try another three digit subtraction problem:

To compute mentally, subtract 300 from 821 to arrive at 521, then add back the complement of 59, which is 41, to arrive at 562, the final answer.

821 – 259 =521 + 41=562

Here is another problem for you to try:

Check your answer and the procedure for solving the problem below;

645 – 372=245 + 28=273

Try the three – digit subtraction exercise given below for practice;

# Chapter 2: Mental Multiplications

In this chapter you'll learn how to multiply in your head one – digit numbers by two digit and three – digit numbers, two – digit numbers by two – digit numbers; two digit numbers by three digit numbers and four – digit numbers by one – digit numbers. Once you have mastered the techniques described in this chapter, you won't have to rely on pencil and paper again. But, there is one prerequisite for mastering the skills in this chapter – you need to know the multiplication table through ten. In fact, to really make headway, you need to know your multiplication tables backward and forward.

2 – BY – 1MULTIPLICATIONS

If you worked your way through chapter 1, you get into the habit of adding and subtracting from left to right. All calculations in this chapter will also be done from left to right as well. (For one thing, you can start to say your answer aloud before you have finished the calculation. That way you seem to be calculating even faster than you are!).

Check this first problem:

×

First multiply 40 × 7 =280. (Note that 40 × 7 is just 4 × 7, with a zero attached). Next multiply 2 × 7 = 14. Then add 280 plus 14 (left to right) to arrive at 294. This is illustrated as follows:

×

40 × 7=280

2 × 7=  14

294

Another example:

×

You first break down the problem into similar multiplication tasks that you can perform mentally with ease.

40 × 4=160

8 × 4=  32

192

Let's do two more!

× ×

60 × 3 =18070 × 9 =630

2 × 3=  68 × 9= 72
186702

 Please, let's try two more; this time a little harder ones

× ×

 40 × 9=36070 × 8=560
8 × 9= 727 × 8= 56
432616
Better still, for those two problems above. I would prefer to round them up i.e. treat 48 as (50 − 2) and 77 as (80 − 3).

× ×

 50 × 9=45080 × 8=640
-2 × 9= 18-3 × 8= 24
432616

The subtraction method works especially well for numbers that are just one digit or two − digits away from a multiple of 10.

So that you can perfect the technique, I strongly recommend practicing more 2 − by − 1 multiplication problems. Below are nine problems for you to tackle. Once you feel confident that you can perform these problems rapidly in your head, you are ready for the next level of mental calculation.

EXERCISE: 2 – BY – 1 MULTIPLICATIONS

1.× 4.× 7.×

2.× 5.× 8.×

3.× 6.× 9.×

3 – BY – 1 MULTIPLICATIONS

Now that you know how to do 2 – by – 1 multiplication problems in your head, you will find that multiplying three – digits by a single digit is not much more difficult. You can get started with the following 3 – by – 1 problem (which is really just a 2 – by – 1 problem in disguise).

×

300 × 7=2100
20 × 7=  140
2240

Was that easy for you? (if not, try practicing more 2 – by – 1 multiplication problems). Let's try another 3 – by – 1 problem similar to the one you just did, except we have replaced the 0 with a 4,so you have another step to perform.

×

300 × 7=2100
20 × 7=  140
4 × 7=    28
2268


Let's try another problem:

×

600 × 4=2400
40 × 4=  160
7 × 4=    28
2588

Even if the numbers are large, the process is just as simple.

 For Example:

×

900 × 9=8100
80 × 9=  720
7 × 9=    63

8883

Let's escalate the challenges by trying a couple of problems that require some carrying.

× ×

100 × 7=  700600 × 9=5400

80 × 7=  560  80 × 9=  720

  12606120

4 × 7=   28  4 × 9=              36

12886156


In the next two problems you need to carry a number at the end of the problem instead of at the beginning.

× ×

600 × 9=5400300 × 4=1200

40 × 9=  36070 × 4=  280

  57601480

8 × 9=   72  6 × 4=   24

58321504

The first part of each of these problems is easy enough to compute mentally. The next two problems would require you to carry two numbers each, so they may take you longer time than the ones you have already done. But with practice you'll get faster.

× ×

400 × 7=2800200 × 9=1800

80 × 7= 56020 × 9= 180

 33601980

9 × 7=  63  4 × 9=  36

34232016

When you are first tackling these problems, repeat the answer to each part out loud as you compute the rest. To reinforce what you have just learned, solve the following 3 – by – 1 multiplication problems in your head.


EXERCISE:3 – BY – 1 MULTIPLICATIONS


1.× 3.× 5.×


2.× 4.× 6.×


2 – BY – 2 MULTIPLICATIONS


The fun of mathematics begins when you realize you can solve a single problem many different ways and still arrive at the answer. The first method you will learn is the "addition method" which can be used to solve all 2 – by – 2 multiplication problems.

The Addition Method

To use the addition method to multiply any two – digit numbers, all you need to do is perform 2 – by – 1 multiplication problems and add the results together. For Example:

×

40 × 46=1840

2 × 46=   92

1932

Here, you split 42 into 40 and 2, two numbers that are easy to multiply. Then you multiply 40 × 46, which is just 4 × 46, with a 0 attached, or 1840. Then you multiply 2 × 46 = 92. Finally, you add 1840 + 92 = 1932 as diagrammed above.

 Here's another way to do the same problem!

×

40 × 42=1680

6 × 42=  252

1932

The catch here is that multiplying 6 × 42 is harder to do than multiplying 2 × 46, as in the first solution. Moreover, adding 1680 + 252 is more difficult than adding 1840 + 92. So in choosing the number to split, I try to choose the number that will produce easier addition problem. In most cases – but not all – you will want to split up the number with the smaller last digit (unit) because it usually produces a smaller second number for you to add.

Now try your hand at this!

x x

 70 × 48=336080 × 59=4720

3 × 48= 1441 × 59=   59

35044779

The last problem illustrates why numbers that end in 1 are especially beautiful to split – up. If both numbers ends in the same digit, you should split up the larger number as illustrated below:

x

 80 × 34=2720

4 × 34= 136

2856

Now you try this in your head:

x

70 × 89=6230

2 × 89= 178

6408

If you manage to get the right answer first or second attempt, celebrate yourself. But before you read on, practice the addition method on the following multiplication problem.

EXERCISE:2 – BY – 2ADDITION METHOD – MULTIPLICATION

1.× 3.× 5.×

2.× 4.× 6.×

The Subtraction Method

The subtraction method really comes in handy when one of the numbers you want to multiply ends in 8 or 9. The following problem explains what I mean:

×

60 × 17=1020

- 1 × 17=   17

1003

Although I concur most people find addition easier than subtraction, it is usually easier to subtract a small number than to add a big number. (If we had done this problem by the addition method, we would have added 850 + 153      = 1003)

 Now let's do the challenging problem from the end of the last section:

×

 90 × 72=6480

- 1 × 72=   72

6408

 Wasn't that easier? Now, here is a problem where one number ends in8.

×

 90 × 23=2070

- 2 × 23=   46

2024

In this case, you treat 88 as 90 − 2, then multiply 90 × 23 = 2070. But you have multiplied by too much. How much? By 2 × 23, or 46 too much. So subtract 46 from 2070 to arrive at 2024, the final answer.

Not only do I use subtraction method with numbers that ends in 8 or 9, but also for numbers in the high 90s because 100 is such a convenient number to multiply. For example, if someone should ask me to multiply 96 × 73, I would immediately round up 96 to 100.

×

100 × 73=7300

- 4 × 73=  292

7008

When the subtraction component of a multiplication problem requires you to borrow a number, using complements can help you arrive at the answer more quickly.

You'll see what I mean as you work your way through the problems below; for example, subtract 340 − 68. We know the answer will be in the 200s. The difference between 40 and 68 is 28. Now take the complement of 28 to get 72. And that's the answer, 272!

68 − 40 = 28

272

Now look at this problem:

×

90 × 76=6840

- 2 × 76= 152

There are two ways to perform the subtraction component of this problem. The "long" way subtracts200 and adds back 48:

6840 – 152= 6640 + 48 = 6688

The short way is to realize that the answer will be 66 hundred and something. To determine the something, we subtract 52 – 40 = 12 and then find the complement of 12, which is 88. So the answer is 6688.

What about this one:

×

60 × 67=4020

- 1 × 67=   67

3953

Again, you can see that the answer will be 3900 and something. Because 67 – 20 = 47, the complement 53 means the answer is 3953.

As you may have realized, you can use this method with any subtraction problem that requires you to borrow a number, not just those that are part of a multiplication problem.

Master this technique and very soon people will be complimenting you.

EXERCISE: 2 – BY – 2 SUBTRACTION METHOD – MULTIPLICATION

1.× 3.× 5.×

2.× 4.× 6.×

## Chapter 3: Squaring Numbers

Squaring numbers in your head i.e. multiplying a number by itself is one of the easiest yet most impressive feats of mental mathematics you can do. In fact, squaring numbers from two digits up to five – digits is what really use to dominate my show. The method I use in squaring numbers is credited to Professor Arthur Benjamin of Harvey Mudd College California. Now let's get started.

Squaring two – digit numbers
Before I show you how to square any two digit numbers, let me show you how to square two digit numbers that ends in 5 first. To square a two-digit number that ends in 5, you need to remember only two (2) things:
1.The answer begins by multiplying the first digit by the next higher digit.
2.The answer always ends in 25

For example to square the number 45, we simply multiply the first digit i.e. 4 by the next higher digit, which is 5, then attach 25. Since 4 × 5 = 20, the answer is 2025.

Therefore, 452 or 45 × 45 = 2025. The steps can be illustrated below:

×

4 × 5=20

5 × 5=   25

2025 – Answer

What of 752? Here we have:

×

7 × 8=56

5 × 5=   25

5625 – Answer


You can try more of that if you will. Now let's move to squaring any two digit number at all. For instance, let's find the square of 23. 23 is not a bad number to multiply but what number is closer to 23 and ends in zero (0), we have 20, so I come down by 3 from 23 to 20. Since I have come down by 3 to 20, I have to balance it by going up by 3 from 23 to 26. So the first part of my calculation is multiplying 26 by 20, which is very easy, you just multiply 26 × 2 and attach a zero (0) to the result here we have 520. Almost done! All you have to add is the square of the number that we went up and down i.e.32 = 9. Adding 9 to 520 gives us 529, which is our answer. The method follows the diagram below:

26

232 26 × 20=520

+ 32=   9

20529

We went up and down by 3. That's the law of gravity. Isn't it? Just kidding!

Now, let's see how this works for another square.

62

612 62 × 60=3720

+ 12=    1

60 3721

80

782 80 × 76=6080

+ 22=   4

766084

60

562 60 × 52=3120

+  42=  16

52        3136

I can tell you, to get very good at squaring two – digit numbers, you need to practice. PRACTICE!Give yourself some pretty tough number to square and using the method you've just learned find the answer. I know that bit by bit you'll soon become very good at it.

Squaring Three – Digit Numbers

Squaring three – digit numbers is another impressive feat of mental calculation. But again, before I show you how to square any three – digit at all, I will again love to show you how to square any three digit number that ends in 25. To square a three digit numbers ending 25, you again need to remember three (3) things:

1.The answer begins by taking the quotient of half the first digit and adding it to the            square of the first digit.

2.The answer always ends in 625.

For example, to square the number 325, we simply half 3 to get 1 remainder 1, we take the quotient 1 and add that to 32, or 9 to get 9 + 1 = 10. To get an answer in the form 10X625.

3.The number X usually varies. But it has two instances to determine it. This           includes:

i.If the first digit of the three digit number is even, X will be 0, that is you put a 0 in the space.

ii.But if the first digit of the three digit number is odd, then X becomes 5, that is you put a 5 in the space.

So for this problem 3252, 3 – the first digit is odd, so the number to be attached will be 5. So we have our answer 105,625.

What of 8252? Half of 8 is 4, add that to 82 or 64 to get 64 + 4 = 68, so our answer will be in the form 68X625, to determine the value of X, look at the first digit, 8 which is even, so X must be 0. This gives us the answer: 8252 = 680, 625.

Let's try one more for a better understanding: suppose we want to find the square of 725. We take half of 7 which is 3 remainder 1, we take the quotient 3, add it to the square of 7, which is 49, so 3 + 49 = 52, from there we look at the first digit 7, it is odd so we attach a 5 to 52 to get the answer 525, 625.

Try solving more problems on that so you can get good at doing it mentally

Now, let's move up to squaring any three digit number at all. Just as you square two – digit numbers by rounding up or down to the nearest multiple of 10, to square three digit numbers, you round up or down to the nearest multiple of 100.

For example, let's square 193:

200

1933200 × 186=37, 200
+ 72=      49
186 37, 249

Further illustration:

712

7062712 × 700=498, 400
+ 62=      36
700498, 436

328

3142328 × 300=98, 400
+ 142=     196
30098, 596
  552


  5262552 × 500=276, 000
+ 262=     676
500276, 676



  900


  8642900 × 828=745, 200
+ 362=   1, 296
828746, 496


Now, do this in your head!9872
  1000


  97421000 × 948=948, 000
+ 262=     676
948948, 676


Why this method works.

As for squaring numbers, the following algebra justifies the method. For any numbers A and d:

$$A^2 = (A + d) \times (A - d) + d^2$$

Here, A is the number being squared, d can be any number, but it is reasonable to choose it to be the distances from A to the nearest multiple of 10.

Hence for $87^2$, I set d = 3 and our formula tells us that:

$$87^2 = (87 + 3) \times (87 - 3) + 3^2 = (90 \times 84) + 9 = 7560 + 9 = 7569$$

Now, having learned how to square two-digit and three-digit numbers, you can escalate the problem to four digit numbers or even five digit numbers if you want to. Let me just use one four digit number as an illustration. Let's square the number 4679. As usual we go up or down to the nearest multiple of 1000, here 5000.

5000

$4679^2$5000 × 4358=21, 790, 000

+      3212=     103, 041

435821, 893, 041

Seems pretty easy, huh? With more practice, you can be best at it.

# Chapter 4: Using Squares To Help Multiply (Shortcuts)

Once you have mastered the art of squaring numbers in your head, you have in your possession a powerful tool applicable to more general problems in multiplication. The product of two numbers that differ from each other only slightly is nearly equal to the square of the number midway between the given numbers. The definite mathematical relationship which exists between the product and the square will be employed in the shortcuts that follow.

1.Multiplying two numbers where difference is1.

Rule: (i) Square the greater number and subtract the number from its squares i.e. if axb and a>b then a×b = a2 − a.

(ii) Square the lesser number and add the number to its squares i.e. if axb and a<b then a×b = a2 + a.

If a>b then a×b = a2 − a 35 × 34 = 352 − 35 = 1225 − 35 = 1190.

If a<b then a×b = a2 + a35 × 36 = 352 + 35 = 1225 + 35 = 1260.

2.Multiplying two numbers whose difference is 2.

Rule:Square the number between the two given numbers and subtract 1.

This shortcut is simple to apply when the square can be found easily. For example suppose we are given to multiply 74 by 76. The number between the two numbers is 75. The square of 75 is quickly found to be 5625.

74 × 76 = 5625 – 1 = 5624 – answer

Multiply 67 by 69

Here the number between the given numbers is 68. Its square is 4, 624.

Therefore; 67 × 69 = 4624 – 1 = 4623

Algebraically, to multiply numbers that differ by 2.

a×b = [(a+b)/2] 2 – 1 67 × 69 = [(67+69)/2]2 – 1

= ]2 – 1

= 682 – 1

= 4624 – 1 = 4623


3.Multiplying two numbers whose difference is 3.

Rule: Square one more than the smaller number and add one less than the smaller number to the result.

For example:Multiply 44 × 47

One more than the smaller number, 44 is 45. Square 45

45 × 45 = 2025

Add one less than 44 to the result.

2025 + 43 = 2068

Therefore: 44 × 47= 2068 – Answer


Algebraically multiplying the numbers that differs by three (3) (a< b)

a×b = (a + 1)2 + (a -1).26 × 29 = (26 + 1)2 + (26 – 1)

= 272 + 25

= 729 + 25 = 754

4.Multiplying two numbers whose difference is 4.

Rule: Square the number midway between the two given numbers and subtract 4

For example: Multiply 69 by 73

The number midway between the two numbers is 71.        Square                71                and subtract 4.        71 × 71 = 5041 – 4 = 5,037

Therefore: 69 × 73 = 5,037


Algebraically, multiplying two numbers that differ by 4.

a × b = [(a+b)/2] 2– 4.64 × 68 = [(64+68)/2]2 – 4

= 662 – 4

= 4356 – 4 = 4352

5.Multiplying two numbers whose difference is 6.

Rule: Square the number midway between the two given numbers and subtract 9

For Example:58 × 64.

The number midway between the two given number is 61. The square of 61 is                found to be 3721, next subtract 9

3721 – 9 = 3712

Again, algebraically multiplying two numbers that differ by 6, we have:

a×b = [(a+b)/2] 2 – 9 51 × 57 = [(51+57)/2]2 – 9

= 542 – 9

= 2916 – 9 = 2907

OTHER MULTIPLYING TECHNIQUES

1.Multiplying by 11

One of my favourite feat of mental math – how to multiply, in your head, any two digit number by eleven (11). You will see how easy it is once you know the secret. Consider the problem:

32 × 11

To solve the problem, simply add the digits, 3 + 2 = 5. Put the 5 between 3 and the 2 and there is your answer.

32 × 11 = 3 (3+2) 2 = 352.

What could be easier? Now you try:

53 × 11 = 5 (5 + 3) 3 = 583

81 × 11 = 8 (8 + 1) 1 = 891.Congratulations.

Now, before you get too excited, I have shown you only half of what you need to                 know.

The double trouble trick of 11x

What if the numbers in the gap add up to a double digit? For example, suppose you want to multiply 98 by 11. So you imagine 9 (9 + 8) 8. But that sum                 in the gap in the middle gives you 9 + 8 = 17, so where do you put those digits? 9(17)8? Easy, just leave the second number (here 7) in the gap as before and                 imagine moving the 1 up a place, so you have (9 + 1) 78 = (10)78.

= 1078.Correct again.


The math behind this is fairly easy if you explore it. Suppose you have the number AB (that's A tens and B units) and you want to multiply by 11. First you multiply by 10. That's easy, 10 × AB =

ABO (A hundreds, B tens, O unit). Then you add another AB, so you get 11 lots of AB altogether, giving you A hundreds, (B + A) tens and (O + B) units i.e. A(B + A)(O + B).

This is exactly what all that sliding numbers around in your imagination has been doing without noticing it. Of course, if the middle (A + B) is more than ten (10) (i.e. it is double digit number), you just slide the first digit up to the hundreds column, and it is sorted!

So can we use the "11 method" to multiply three – digit numbers or larger by eleven? Absolutely!
For example, for the problem 314 × 11, the answer still begins with 3 and            ends with 4.
Since 3 + 1 = 4 and 1 + 4 = 5, the answer is 3454
The math behind this is:
ABC × 11 = A (A + B) (B + C) C.

2.Multiplying by other two digit numbers ending in 1 (Y = 1 to 9).
a × Y1 = a + Y0a.63 × 41 = 63 + (40 × 63)
= 63 + 2520 = 2583

3.Multiplying with numbers ending in 5 (Y = 1 to 9).

a × Y5 = a/2 × 2 (Y5)83 × 45 = 83/2 × 2 (45)

= 41.5 × 90

= 415 × 9 = 3735


4.Multiplying by 15

a × 15 = (a + a/2) × 1077 × 15 = (77 + 77/2) × 10

= (77 + 38.5) × 10

= 1155

5.Multiplying by 45

a × 45 = 50a -  59 × 45 = 50 (59) −

= 2950 − 295 = 2655

6.Multiplying by 55

a × 55 = 50a -  67 × 55 = 50 (67) −

= 3350 − 335 = 3685


7.Multiplying by two digit numbers that end in 9 (Y = 1 to 9).

a × Y9 = (Y9 + 1) a − a 47 × 29 = (29 + 1) 47 − 47

= 30 (47) − 47

= 1410 − 47 = 1363


8.Multiplying by multiples of 9 (b = multiple of 9 up to 9 × 9) a×b = round up to next highest 0 multiply 29 × 54

= 29 × 60 − (29 × 60) ×

then subtract   of result.= 1740 − 174 = 1566

# Chapter 5: Rooting Numbers

Simplified Square Roots:

Square roots can also be calculated easily if you are given a perfect square. For instance, if someone told you that the square of a two – digit number was 7569, you could immediately tell her that the original number (the square root) is 87. Here's how! First you have to memorize the squares of number 1 through 10.

12 – 1

22 – 4

32 – 9

42 – 16

52 – 25

62 – 36

72 – 49

82 – 64

92 – 81

102 – 100.

To find the square root of 7569:

i.Look at the magnitude of the "hundreds number" (the number preceding the last two –        digits), 75 in this example.

ii.Since 75 lies between 82 (8 × 8 = 64) and 92 (9 × 9 = 81), then we know that the square root lies in the 80s. Hence, the first digit of the square root is 8. Now there are two numbers whose square ends in 9. 32 = 9 and 72 = 49. So the last digit must be 3 or 7. Hence, the square root is either 83 or 87. Which one?

iii.Compare the original number with the square of 85 (which can be easily computed as 80 × 90 + 25 = 7225). Since 7569 is larger than 7225, the square root is the larger number, 87.

Let's do one more example. What is the square root of 4761?

Since47 lies between 62 = 36 and 72 = 49, the answer must be in the 60s. Since the last digit is 1, the last digit of the square root must be 1 or 9. Since 4761 is greater than 652 = 4225, the square root must be 69.

Note: This method can only be applied when the original number given is a perfect square.

QUICK CUBE ROOTS

1.Cube roots of two – digit number cubes

Ask someone to select a two – digit number and keep it secret. Then have him cube the number, that is, multiply it by itself thrice (using a calculator). For instance, if the secret number is 68, have the volunteer compute 68 × 68 × 68 = 314, 432. Then ask the volunteer to tell you his answer. Once he tells you the answer (i.e. the cube), 314, 432, you can instantly reveal the original secret number, the cube root, 68. How do you do that? Well I'm glad you asked. To calculate cube roots, you need to learn the cubes from 1 – 10.

13 – 1

23 – 8

33 – 27

43 – 64

53 – 125

63 – 216

73 – 343

83 – 512

93 – 729

103 – 1000.

Once you have memorized these, calculating cube roots is as easy as ABC. For instance, what is the cube root of 314, 432?

i.Look at the magnitude of the thousands number (the number left of the comma), 314 in this case.

ii.Since 314 lies between 63 = 216 and 73 = 343, the cube root lies in the 60s            (Since 603 = 216, 000 and 703 = 343, 000). Hence, the first digit of the cube root            is 6.

iii.To determine the last digit of the cube root, note that only the number 8 has a cube that ends in 2 (83 = 512), so the last digit ends in 8.

Therefore, the cube root of 314, 432 is 68. Three simple steps and you are there.


Notice that: Every digit, 0 through 9, appears once the last digits of the cubes. (In            fact the last            digit of the cube root is equal to the last digit of the cube of the last digit of the            cube). Go figure out that one!


Now you try the following practice.

What is the cube root of 19, 683?

i.19 lies between 8 and 27 (23and 33)

ii.Therefore the cube root is 20– something

iii.The last digit of the problem is 3, which corresponds to 343 = 73. So 7 is the last digit         of the cube root.

OR

iii.The last digit of the problem is 3, 33 = 27. So 7 is the last digit of the cube root.

The answer is 27.


Also Note that: Our derivation of the last digit will only work if the original number is the cube of a whole number.


2.Cube roots of three – digit number cubes

Finding the cube roots of two – digit number cubes is impressive but finding the cube roots of three – digit number cubes is much more impressive and outstanding.


Ask someone to cube a three – digit number, and ask how many digits are in the answer (it should be seven, eight, or nine digits). Ask the volunteer to recite (or            write down) the answer and you can pretty easily determine the cube root in your head. The first digit and last digit of the cube root are determined exactly as             before.

The last digit of the cube root is uniquely determined by the last digit of the cube. (In fact, the last digit of the cube root is the last digit of the cube of the last digit of the cube!). For the first digit, we look at the magnitude of millions. But the real work comes, when it is time to determine the middle digit. Basically, there are two methods of determining the middle digit, they are:

i.Casting out nines, or

ii.Casting out elevens.

But I'll only teach you the casting out elevens because the casting out of nines would still require you to do a little guess work to get the middle digit and you may even be wrong if you are not careful, but with a little extra mental effort and memory, by using the casting out elevens we can determine the middle digit without any guess work, because the cubes of the numbers0 through 10 are all distinct mod 11, as shown below;

Table of Cubes mod 11

| n | 0 | 1 | 2 | 3 | 4 | 5 |
|---|---|---|---|---|---|---|
|   | 6 | 7 | 8 | 9 | 10 |   |
| n3 | 0 | 1 | 8 | 5 | 9 | 4 |
|   | 7 | 2 | 6 | 3 | 10 |   |

For Example:

For the cube root of 19, 248, 832:The first digit of the cube root must be 2 (since 19 lies between 23 = 8 and 33 = 27) and the last digit of the cube root must be 8 (since the cube ends in 2 and 23 = 8, so we take the 8 in the end). Hence the answer must be in the form 2 8. How do we determine the middle digit? By alternately adding and subtracting its digit from right to left, we see that it reduces to 2 − 3 + 8 − 8 + 4 − 2 + 9 − 1 = 9 (Mod 11). Hence the cube root must reduce to 4 (mod 11). Since the answer is in the form 2 8, then 2 + 8 −  = 4 (mod 11) results in and answer of 268.

Another Example:

To find the number with cube 111, 980, 168. We see immediately that the answer will be in the form 4 2. Casting out 11s from left to right, we compute $1 - 1 + 1 - 9 + 8 - 0 + 1 - 6 + 8 = 3$ (mod 11). Here, the cube root must reduce to 9 (mod11). Thus $4 + 2 - = 9$ (mod 11) tells us that the missing digit is 8. Hence the cube root is 482.

Note that:

i.If the cube has 7 or 9 digit, then you can alternately add and subtract the digits in their natural order.

ii.If the cube has 8 digits, you can do it in two ways:

- You can start with 11, then alternately subtract and add the subsequent digits e.g 19, 248, 832.

$=11 - 1 + 9 - 2 + 4 - 8 + 8 - 3 + 2 = 9$ (mod 11).

- You can also do it by alternately subtracting and adding its digits from right to left (as we did in the example above) e.g. 19, 248, 832 $= 2 - 3 + 8 - 8 + 4 - 2 + 9 - 1 = 9$ (mod 11).

iii. But what if the process results in a negative mod as in: 2613 = 17, 779, 581.

By alternately adding and subtracting from right to left we have:

$1 - 8 + 5 - 9 + 7 - 7 + 7 - 1 = -5$ (mod 11).

All you need to do is to add 11 to the negative mod - 5 + 11 = 6 (mod 11). So the cube root must reduce to 8 (mod 11).

If 2 1 = 8 (mod 11)

Then 2 −  + 1 = 8 (mod 11)

2 + 1 −  = 8 (mod 11) results in an answer of 261.


One Last example:

Let's find the cube root of the perfect cube 377, 933, 067. The first digit of the cube root must be 7 (since 377 lies between 73 = 343 and 83 = 512) and the last digit of the cube root must be 3 (since the cube ends in 7 and 73 = 343, so we take the 3 in the end). Hence, the cube root must be in the form 7 3. Since the cube has 9 digits, we alternately add and subtract its digit from left to right to get: $3 - 7 + 7 - 9 + 3 - 3 + 0 - 6 + 7 = -5$ (mod 11) or 6 (mod 11). Which means the cube root must reduce to 8 (mod 11).

So, we have 7 3 = 8 (mod 11)

7 −  + 3 = 8 (mod 11)

7 + 3 −  = 8 (mod 11)

Which results in an answer of 723.Bravo!


QUICK 5TH ROOTS


5th roots of two – digit numbers 5th power

Taking the 5th root is even easier than the cube root. Because, you have 5 =  (mod 10), so the last digit is simple. For the first digit you memorize the following table:

Size of  5

1100 thousand

23 million

324 million

4100 million

5300 million

6777 million

71. 6 billion

83 billion

96 billion


Ask someone to type a two digit number (in a calculator) and raise it to the power of 5 (or multiply it by itself 5 times). Now ask the volunteer to tell you his answer. Let's say you hear 8,587,340,257. As soon as you hear 8 billion, you know by the table that the first digit must be 9. You ignore the rest and wait for the last digit which happens to be 7, so the result is 97.

Another example:

What if the volunteer call out 992,436,543? As soon as you hear 992 million, you know by the table that the first digit has to be 6. You ignore the rest of the digits being called out and wait for the last digit which is 3. So the result is 63. Cool! Isn't it"?

## Chapter 6: Subtraction

Mental subtraction is one of the most often needed and used skills in our daily life. In the shop, at work, or at home, the ability to quickly subtract 2-digit, 3-digit, 4-digit numbers or higher is handy. And the beautiful thing about most mental tricks and methods for subtraction is that they allow you to subtract even large 6-digit and 7-digit numbers just as easily as the smaller ones. They take more practice, of course, but once you've gotten the hang of it, you will surprise yourself – and everyone around you – with your skills.

Another great thing about getting good at mental subtraction of bigger numbers is that it trains several different mental skills. It improves your calculus, your memory, as well as your ability to keep several things in focus and to multitask. This, in turn, can help you a lot of with complex mental math problems, as well as with a ton of daily activities.

So, without further ado, let's go over several different subtraction methods:

Subtract in two or more distinct steps.

Some big subtractions can seem daunting at first – such big numbers, so many digits. That's why it's often helpful to divide the subtracting process into several smaller steps. Let's start with small numbers:

57 – 9 =

Now, this is an easy example that doesn't need more than one calculation, but for the sake of the example, let's divide it into two steps:

57 – 7 – 2 = 48.

Because we know that 9 = 7 + 2, we can first subtract the 7 from 57 to get it down to a nice, round number like 50, and then subtract the other 2 to find the answer – 48.

The same method can be used for bigger numbers:

239 – 84 =

First, we can subtract 39 from 84, and we will get 45. This means that 84 = 39 + 45. With that in mind, we can finish the initial subtraction like this:

239 – 39 = 200

200 – 45 = 155

It feels much simpler this way, doesn't it? Of course, if you are out of practice, you may find it difficult to do this in your mind at first, but after some practice with smaller numbers, you'll quickly move on to 3-digit numbers and above.

Let's try the same method with 4-digit numbers:

3856 – 1745 =

Using this method, this subtraction will take several steps. First:

1745 – 856 =

856 – 745 =

745 – 56 =

56 – 45 = 11

So: 745 – 56 = 745 – 45 – 11 = 700 – 11 = 689

856 – 745 = 856 – 56 – 689 = 800 – 689 = 111

1745 – 856 = 1745 – 745 – 111 = 1000 – 111 = 889

3856 – 1745 = 3000 – 889 = 2111

It is fun, isn't it? Now, for 4-digit numbers and above, this isn't the fastest method of mental subtraction. The multiple different steps can be hard to keep track of with so many numbers and digits. However, this is a great training method, as it requires you to not only perform multiple subtractions, but it asks you to train your memory and to multitask.

For a more practical method:

Subtracting by dividing each number into its components.

Here is another multi-step method that doesn't become (much) more complicated at higher digit numbers. It's a much more practical method for mental calculation which you can use in a pinch, but it isn't as great of a training exercise as the previous method.

Again, let's start with a small number:

99 − 28 =

If we break these two numbers into their components, we'll get:

(90 + 9) − (20 + 8) =

(90 − 20) + (9 − 8) =

70 + 1 = 71

This method utilizes the fact that each number consists of the sum of its parts, which is best viewed by breaking them into tens. 99 consists of 9 tens and one 9. And 28 consists of 2 tens and one 8. Subtracting 2 tens from 9 tens is easy, as is subtracting 8 from 9.

What happens if the smaller number ends on a bigger digit than the big number, however? Let's check out a similar 2-digit example:

98 − 29 =

If we break the two numbers the same way, we may face a problem:

$(90 + 8) - (20 + 9) =$

$(90 - 20) + (8 - 9) =$

$70 + (-1) = 70 - 1 = 69$

If you're comfortable with negative numbers than you'll have no problem with this method – whenever you need to add a negative number you subtract it instead.

If this feels counter-intuitive or uncomfortable to do mentally, however, here's another solution:

$98 - 29 =$

$(80 + 18) - (20 + 9) =$

$(80 - 20) + (18 - 9) =$

$60 + 9 = 69$

The result is the same, and the principle is the same, only we moved one-tenth from the 9 tens to the right-most part of the number to ease the next subtraction.

Let's try it with some bigger numbers:

$853 - 186 =$

$(800 + 50 + 3) - (100 + 80 + 6) =$

$(800 - 100) + (50 - 80) + (3 - 6) =$

$700 + (-30) + (-3) = 700 - 30 - 3 = 667$

Or, to do it without negative numbers:

853 − 186 =

(700 + 150 + 3) − (100 + 80 + 6) =

(700 + 140 + 13) − (100 + 80 + 6) =

(700 − 100) + (140 − 80) + (13 − 6) =

600 + 60 + 7 = 667

Both methods are quite easy with 3-digit numbers, it's just a matter of which one you prefer. Let's see a 4-digit example too:

3542 − 1674 =

(3000 + 500 + 40 + 2) − (1000 + 600 + 70 + 4) =

(3000 − 1000) + (500 − 600) + (40 − 70) + (2 − 4) =

2000 + (-100) + (-30) + (-2) = 2000 − 100 − 30 − 2 =

1900 − 30 − 2 = 1870 − 2 = 1868

Or:

3542 − 1674 =

(2800 + 700 + 40 + 2) - (1000 + 600 + 70 + 4) =

(2800 + 600 + 140 + 2) - (1000 + 600 + 70 + 4) =

(2800 + 600 + 130 + 12) - (1000 + 600 + 70 + 4) =

(2800 − 1000) + (600 − 600) + (130 − 70) + (12 − 4) =

1800 + 0 + 60 + 8 = 1868

Rounding one or both numbers up or down to ease the subtraction.

Otherwise known as The Golden Rule of subtraction, this easy trick makes most mental subtractions rather simple. Here's how it works with 2-digit numbers first:

54 – 38 =

This is an annoying little subtraction at first glance, but if we change it to this:

54 – (40 – 2) =

(54 – 40) + 2 =

It suddenly becomes much easier.

54 – 40 = 14

14 + 2 – 16

All we did was round up 38 to 40. The only tricky part is to remember to count for the 2.

Here's the same principle with bigger numbers:

342 – 167 =

(340 + 2) – (170 – 3) =

(340 – 170) + (2 + 3) =

170 + 5 = 175

As you can see, this is a bit trickier because we rounded both numbers. What's more, we rounded 342 down to 340, and we rounded 167 up to 170. This makes the addition of the 2 and the subtraction of the 3 a bit counterintuitive –

you might ask we the second half of the third step looks like this:

+ (2 + 3)

Instead of this:

+ (2 − 3)

The reason is that, in the second step of the subtraction, there is a double negative in front of the 3:

- (170 − 3)

And a double negative equals a positive, so we add both the 2 and the 3 to 170 to get 175.

Let's try with 4-digit numbers, shall we?

3247 − 2359 =

(3250 − 3) − (2400 − 41) =

(3250 − 2400) + (-3 + 41) =

850 + 38 = 888

Simple, right? The beautiful thing about this method is that you can round numbers up and down to whatever you feel like it. Is 3250 not around enough number? Round 3247 to 3200 then − it will look like this:

3247 − 2359 =

(3200 + 47) − (2400 − 41) =

(3200 − 2400) + (47 + 41) =

800 + 88 = 888

The only tricky thing about this method is understanding when you should add and when you should subtract the add-on numbers. Once you get used to remembering that a double negative = a positive, however, the Golden Rule becomes, well – golden.

Some practice tests.

Let's, for practice's sake, do some higher digit numbers with our 3 methods and see what happens. We won't solve these for you, so you can try and solve them yourself. As they are more complex, you don't need to try and do them in your mind at first – use the good old fashioned pen & paper to practice. For the best results, try and use all 3 methods on each of the following subtractions. Start with the step-by-step method, as while it's slower, it gives you the most practical and trains you. Then try doing all the examples by dividing the numbers into their components, and in the end – try the Golden Rule.

1.  17 589 – 9621 =?
2.  32 325 – 13 497 =?
3.  21 568 – 7899 =?
4.  63 144 – 35 651 =?

5.   127 349 – 51 568 =?


Multiplication

Mental multiplication is one step more complicated than division, but several different methods can make it simpler. And just as with division, the ability to multiply at least 2-digit and 3-digit numbers in our minds is very handy in our day-to-day lives. It's very helpful in most professions; it's helpful in dealing with taxes faster, it is useful in calculating interests and return investments, and so on. Not to mention that learning to multiply 4-digit, 5-digit or higher numbers in your mind is a surefire way to very effectively train your memory, your mental visualization, your multitasking abilities, and so on. Simply put, becoming a human calculator puts you one step closer to becoming a human computer.

So, what should we do if we want to multiply significant numbers by ourselves? Well, before we go to the several methods we have for your below, let's see what the basics you need to cover first look like:

- You need to know the multiplication table in your sleep. Any single-digit multiplication between 1 x 1 and 9 x 9 should be intuitive for you. 6 x 7? 9 x 5? 3 x 8? If any of these are giving you pause and need to think about them for even a fraction of a second – take the single-digit multiplication table and memorize it. Answering those multiplications should be as easy as associations like "summer = warm" and "winter = cold."

- As a direct consequence, you should also have no problem multiplying round numbers that end on zeroes. 60 x 70 is no different than 6 x 7; you need to add 2 zeroes in the end. 900 x 50 is also no different than 9 x 5; you need to add 3 zeroes this time.

- You need to know how to multiply big (at least 3-digit and 4-digit) numbers with a pen and paper. The traditional right-to-left method that we learn in school is not something that you should try and do mentally – it's too slow and clumsy. But, you should feel perfectly comfortable with it on paper if you are to try the mental methods we have for you below.

• You should know the difference between a multiplicand and multiplier. Simply put, the multiplicand is the number that is being multiplied, and the multiplier is the number by which you multiply the multiplicand. Of course, these are interchangeable – in 3 x 2 you can multiply 3 by 2, or you can multiply 2 by 3. Typically, the bigger number is the multiplicand, and the smaller number is the multiplier since this tends to be easier. Regardless of your preferences, however, you should be aware of both terms.

So, with all that in mind, here are some mental methods for multiplying:

General left-to-right method.

As we mentioned, when we're multiplying by hand in school we typically multiply from right to left. This is done so that we can be sure not to miss anything, as well as so that the students can understand and learn the entire process. When multiplying mentally, however, this is a somewhat abstract method as it takes a lot of time. Instead, multiplying from left to right requires a bit more thinking, but is much faster and more intuitive.

That's why we said that you need to have mastered the right-to-left method on paper first – so that you understand the principles of multiplication. This way, with enough mental exercises you'll multiply big numbers from left to right in your head with superior speed. What's more, after enough practice you'll learn to accurately guess the answer before you've finished multiplying the individual digits.

So, how does this method work? Simple – you multiply digit by digit, going from the left to right. Here are several examples (we'll start with a single-digit multiplier):

7359 x 6 =

7000 x 6 + 300 x 6 + 50 x 6 + 9 x 6 =

42 000 + 1800 + 300 + 54 = 44 154

Now, this seems a bit complicated to do in your mind – there are so many zeroes! But if you ignore the zeroes and remember the positions of the numbers you can go through the following mental process:

7000 + 6 is 42 000, so the result will be 5-digit (so, it will have 5 "positions" for its digits) and will likely be between 42 000 and 50 000. So:

7 x 6 = 42 in the first 2 positions of the number

3 x 6 = 18, and the ten goes to the 42, making it 43, leaving 8 in the 3rd position of the number

5 x 6 = 30 in the 4th position, but the 3 goes to the 8, making it 11, which makes it 1, with the 10 going to the 43 making it 44. In the 4th position, we are left with a 0 for now

9 x 6 = 54 in the last two positions of the number. So, that's 44 1 54.

Simple, right? Well, it looks like an awful lot of explanation, so let's look at it this way:

```
  7359
x    4
42
  18
   30
    54
=44154
```

Now it looks familiar, right? Just as the right-to-left method we did in school, but in the opposite direction. It requires a bit more memory to do in your mind, which is why they thought is the other way – so we don't make mistakes. Once you've become proficient enough, however, you won't make mistakes, and you'll do it much quicker.

Rounding up the multiplicand.

Just as we did with the subtraction, we can also round up numbers when we're multiplying them. Again, it requires a bit of memory, so you don't forget what number you added or subtracted, but once you've got that, the rest is easy. Let's try with a single-digit multiplier again:

59 x 6 =

This becomes much easier if we round it to:

60 x 6 = 360

However, since we rounded up, we should also subtract 1 x 6 = 6 and:

360 − 6 = 354

Let's try with a bigger number:

4359 x 8 =

This can easily be switched to:

4500 x 8 − 8 x 141 (the difference between 4500 and 4359) =

36000 − 1128 = 34872

And you can calculate 141 x 8 in the very same way:

141 x 8 =

150 x 8 - 9 x 8 =

1200 − 72 = 1128

As you can see, the higher the numbers get, the more steps this method has. This makes it feel harder, but with a bit of memory work, it isn't. And if that bit of memory works pose a problem at first, that's all the more reason to keep practicing this method.

Break the multiplier when multiplying two bigger numbers.

Multiplying feels easier when one of the numbers consists of just one digit. Sure, the other number can be enormous, and this can lead to a lot of calculations, but at least it's manageable, right? What happens when neither the multiplicand nor the multiplier is single-digits, however? Well, one simple method is to break one of them into manageable digits:

46 x 42 =

46 x 40 + 46 x 2 =

1840 + 92 = 1932

(And if 46 x 40 seems like it might be trouble, remember that it's no different from 46 x 4, but with a 0 after it. Or, you can also calculate it as 40 x 40 + 6 x 40 = 1600 + 240 = 1840)

Now, this is a surprisingly quick equation, isn't it? Let's try a bigger one:

572 x 29 =

572 x 20 + 572 x 9 =

Using either of the previous two methods, both of these equations are easy.

572 x 20 = 11 440

572 x 9 = 5148

11 440 + 5148 = 16 588

A piece of cake. Let's try another one:

349 x 121 =

Looks intimidating, but it's quite easy.

349 x 100 + 349 x 21 =

34 900 + 349 x 20 + 349 x 1 =

35 249 + 3490 x 2 =

35 249 + 6980 = 42 229

The bigger the numbers are, the more times you'll have to break them. That's why it's easier to break the multiplier instead of the multiplicand since it's usually the smaller number. However, depending on the exact figures, it might be easier to break the bigger multiplicand, if it's a simpler number.

Alternatively, from the last example, you might have figured out the next method:

Rounding up the multiplicand or the multiplier.

Just as we did with single-digit multipliers, rounding up one of the numbers can make the whole equation much easier to calculate mentally. In the previous example, the multiplicand was 349 which is very easy to round up to 350. Then the equation would have become 350 x 121 − 1 x 121 = 42 350 − 121 = 42 229. But let's try with another example here:

487 x 136 =

500 x 136 − 13 x 136 =

68 000 − 13 x 136 =

68 000 − 10 x 136 − 3 x 136 =

68 000 − 1360 − 408 = 66 232

(If 500 x 136 gave you trouble, remember that you can do it like this: 500 x 136 = (1000 x 136) / 2 = 136 000 / 2 = 68 000)

As we can see, rounding numbers makes mental math exceptionally easy − all that's required is a little bit of memory so that you don't forget how much you rounded one of the numbers with.

The factoring method.

This method may feel a bit harder at first, but it makes the whole process much faster and easier once you understand it. It removes the need to remember too many numbers and equations. The basis of the factoring method is that instead of rounding up the multiplier or the multiplicand, you divide one of them into suitable numbers. Now, this method doesn't work for all numbers, since not all numbers divide comfortably, especially when you're doing it in your mind. But still, it works for a lot of numbers. Here's what it looks like:

63 x 21 =

63 x (7 x 3) =

(63 x 7) x 3 =

(60 x 7 + 3 x 7) x 3 =

(420 + 21) x 3 =

441 x 3 = 1323

As you can see, this method takes advantage of the fact that 21 = 7 x 3. Alternatively, you could have taken advantage of the fact that 63 = 7 x 9, but those are a bit higher numbers, so 7 and 3 are more comfortable to use. You can use this method to calculate very high numbers, as long as one of them breaks up into easy-to-use numbers. Let's try this one on for size:

459 x 35 =

459 x (5 x 7) =

(459 x 7) x 5 =

From this point on, it's easy.

3213 x 5 =

(3213 x 10) / 2 =

32 130 / 2 = 16 065


Some practice tests.

As you can see, mental multiplication is a bit more complicated than subtraction, but by mastering these basic tricks, it can become easy enough to do on the fly. It requires a bit of memory, as well as managing to keep your focus on several numbers at the same time, which can be daunting at first. With enough practice, however, it will quickly stop being too complicated. And speaking of practice – there are several harder examples to play with. Use a pen & paper at first, but don't slack and try different methods of each example:

29 x 47 =

321 x 94 =

586 x 77 =

133 x 386 =

454 x 638 =

3298 x 542 =

Division

Just like multiplication, the division can be a bit trickier than subtraction and addition, but it's nothing unmanageable once you've gotten the hang of it. There are multiple clever ways to easily divide big numbers mentally without the need for calculators or paper. They all require a bit of memory work sure, but practicing them not only will make them easier – it will also easily improve your memory skills too. What's more mental division itself is exceptionally useful for calculating percentages, interest rates and other statistics on the fly without wasting time with a calculator or making a fool of yourself.

There are, of course, several things that need to be kept in mind before we can begin. As with mental multiplication, you need to make sure that you've got the basics down:

• You need to be able to multiply and divide basic numbers without a hitch. Everything from the multiplication table should be as simple for you like the letters of the alphabet. 81 / 9? 42 / 7? Simple divisions up to a 100 should come naturally for you. If that's not the case, get the multiplication table and start practicing – even just memorizing that can be of great help when dealing with percentages, which alone has a ton of practical uses in our day to day life.

•      Dividing numbers that divide to 10 should be no issue for you as well. 140 / 7? It's no different than 14 / 7; it just has a zero at its end. 630 / 90? Simple, it's no different from 63 / 9.

•      Division with big numbers (3- or 4-digit numbers at least) should also be no problem or you if you have a pen and paper in front of you. Mental division utilizes other methods than normal, "on paper" division, but knowing the basic way is a requirement. Without understanding that, you'll have a hard time learning how to do mental division.

•      The difference between a dividend and a divider should be clear before we start as well. The dividend is the number you divide – the first number in the equation. It's usually bigger, but that's not a necessity. The divider is the number you divide the dividend too. It's the number after the division sign, and it's usually smaller than the dividend.

•      You should know from the first glance whether a big number divides into 2, 3 or 5. This is simple for 2 and 5, but it's trickier for 3. The knack is that if the sum of the digits of a number divides into 3, then the number itself divides into 3. For example, we know that 168 divides into 3 because 1 + 6 + 8 = 15 and 15 divides into 3. It's important to remember that for the mental division so you can simplify a lot of the equations, you'll face.

• You should keep in mind that when it comes to division, the result is often not a whole number. The result of a multiplication equation is always a whole number as long as the multiplicand, and the multiplier is whole as well. That's not so with division – you can take two whole numbers, and the result of their division won't be a whole number. 9 / 3 = 3, but 9 / 2 = 4.5 and 9 / 5 = 1.8. Before you start practicing mental division, you should make sure that you are comfortable with such numbers.

Now that we've gotten the basics down, let's start with some simple tips and methods:

Estimate if possible.

The first point here is not so much a method for the mental division as it is a simple trick. Whenever you have division equation in front of you, you should always check out if you can accurately estimate the answer before trying anything else. You'd be surprised at how easy it is sometimes. Let's look at some examples:

159 / 5 =

159 doesn't divide precisely into 5 because it doesn't end on 5. However, 160 divides into 5 for sure, because it divides into 10. So, if we divide 160 into 5, we'll get 32, and we'll know that the result is slightly less than that. In most cases in real life, the exact number is not vital, unless you're calculating something important. If that's the case, you can divide 155 into 5 instead, and you'll get 31. Add the 4 /5 that you ignored to get from 159 to 155, and the result will be a simple 31.8.

This method can be repeated with even bigger dividends and dividers, as long as the numbers make it comfortable and easy to use. Let's see this example:

569 / 25 =

Once again, 569 doesn't divide precisely into either 25 or 5. However, 550 divides into 25, and it yields the simples answer of 22.

So, the answer of 569 / 25 = 22 + 19 / 25, or 22.76

Of course, you'll have to approximate the 0.76, but once you get enough experience, you'll start pinpointing even those numbers with ease.

Dividing with even numbers.

In math as in life, one should always try to simplify things whenever possible. Of course, this isn't always possible – some things are just too darn complicated – but we should always look for ways to simplify them anyway. When it comes to division, that's rather simple. Whenever the dividend and the divider are even numbers, you can always simplify them. Here are some examples:

360 / 40 =

180 / 20 =

90 / 10 = 9

Of course, in this particular case you can also scratch the zero:

360 / 40 =

36 / 4 = 9

Here's another example:

975 / 15 =

195 / 3 = 65

In this case, it's obvious that both numbers divide by 5, so you can do that instead of trying to divide 975 by 15 directly.

Here's another example:

2736 / 144 =

In this case, we can see that both numbers divide into 3. This is clear because the sum of each number's digits divides by 3. 2 + 7 + 3 + 6 = 18, which divides into 3 and 1 + 4 + 4 = 9 which also divides into 3. So:

2736 / 144 =

912 / 48 =

Once again, these numbers divide into 3, so:

304 / 16 =

Here, both numbers divide into 4:

76 / 4 =

38 / 2 = 19

As you can see, without having to use any sophisticated formulas or without using paper or a calculator, we can easily calculate that 2736 / 144 = 19 in a matter of seconds.

Splitting the dividend.

For a more sophisticated method, here's how you can solve mental division by splitting the dividend. Just as we did with mental multiplication, here you can simply split one of the numbers into more comfortable numbers. Let's see some examples:

79 / 7 =

70 / 7 + 9 / 7 =

10 + 1.28 = 11.28

Or for something more complicated:

358 / 13 =

260 / 13 + 98 / 13 =

20 + 7.53 = 27.53

If you're wondering how to quickly divide 98 into 13, this is obviously a tricky question since 13 is a prime number and it doesn't divide into anything but 1 and itself. However, it's also a simple 2-digit number, so you can just quickly multiply 13 several times to see when it gets close to 98. 13 x 2 is 26, 26 x 2 is 52, 52 x 2 is 104. 104 = 13 x 8 and is pretty close to 98, so 98 / 13 is obviously around 7.5

Here's another example:

1947 / 45 =

1800 / 45 + 147 / 45 =

600 / 15 + 147 / 45 =

120 / 3 + 147 / 45 =

40 + 135 / 45 + 12 / 45 =

43 + 4 / 15 = 43.26

In this examples we did several things:

First, we split the dividend into a number that we know will divide easily with 45 because both 1800 and 45 divide into 3 and 15, and into another number (147) that's a bit more uncomfortable but is small enough to be easy.

Next, we simplified 1800 / 45 several times to get it down to 120 / 3 which is an easy 40.

Then we split the 147 dividends into two more numbers – 135 and 12. We chose 135 because we can easily see that it can be divided by 45. 45 x 2 is 90 and 90 + 45 is 135.

The last part was dividing 12 by 45 which is uncomfortable because it doesn't split into a whole number, but is eased by the fact that both numbers split into 3. 4 / 15 is awfully similar to 4 / 16 which is 0.25, so even without a calculator, we can see that 4 / 15 would be 0.26.

Splitting the divider into factors.

Aside from splitting the dividend, we can also split the divider. Instead of splitting that into its components, however, we'd do much better to split it into its factors. Once this is done, we just have to divide the dividend into the factors of the divider. Here are some examples:

816 / 6 =

816 / 2 / 3 =

408 / 3 = 136

Simple, right? Let's try with bigger numbers:

1916 / 15 =

1916 / 3 / 5 =

(1800 / 3 + 116 / 3) / 5 =

(600 + 111 / 3 + 5 / 3) / 5 =

(637 + 1.66) / 5 =

638.66 / 5 =

635 / 5 + 3.66 / 5 =

127 + 7.32 / 10 =

127 + 0.732 = 127.732

Once again, here we did several steps, each of which is simple enough to do mentally. First we split the divider into factors, than we split the dividends into its components. After that, we even split the second, smaller dividend (116) into its components as well. With those actions we got to 638.66 / 5 and we split that dividend into comfortable components as well. The remaining 3.66 / 5 looked a bit irksome, but by doubling it we only needed to divide it into 10, which is always easy.

And that was it. A method such as this one required multiple actions and can be troublesome if you haven't practiced your memory for a while. That is easily fixable with a bit of practice, however.

Some practice tests.

As you can see, the mental division is a bit trickier than subtraction and multiplication. It is less intuitive, and it sometimes requires more memory work. What's even more annoying is that it often doesn't lead to a whole number as a result. All those problems are easily overcome with practice, however, not to mention that sometimes mental division can be deceptively simple when the dividends and the divider are even or comfortable enough.

Either way, if you practice enough, you'll quickly become a master human calculator, capable of the mental division of even bigger numbers. To further help you with that, here are several more practical examples. They are more complicated than the previous ones and require more memory work and calculations. Since you're just starting, don't hesitate to use a piece of paper at first, as long as you are using the methods we outlined above. Once you're done, check the results with a calculator. If something isn't right, go step by step through your calculations and find your mistake – there's no need to be annoyed by mistakes, they are a natural part of the process. Once you get the hang of these examples, start practicing in your mind with other numbers until you master mental division.

834 / 27 =

358 / 65 =

1394 / 113 =

3875 / 248 =

9457 / 623 =

Memory and how to improve it

Most methods we practiced above required quite a bit of memory work to be done mentally. In truth, the mental work that's required is often what turns people off math and mental math in particular – when you're out of practice, it just seems way too daunting.

Memory is very easy to improve, however. We are used to viewing our mind's current state as its "normal" state, but this isn't true. Our mental faculties are very malleable and easy to change, with memory being a prime example for that.

There are a lot of different ways to improve your memory, especially if you try to Google them. There are countless of different paid courses and "mental games" online, all of which promise great results, but only a few of which are worth their money.

If you want to train your memory, you'd usually do much better if you go to the basics:

Eat right.

It's probably annoying to constantly see this as the first tip for virtually every problem, but that's simply the reality of our bodies. The things we eat don't just affect our mass accumulation and our hearts – they affect our entire bodies, including our brains. Sugars and carbohydrates have an adverse effect on our mental prowess, while fresh vegetables and healthy fats do wonders for our brains. Things like curry, celery, broccoli, cauliflower, walnuts, and many others, contain antioxidants and other components that protect our brains and stimulate the production of new brain cells.

Focusing on animal-based omega-3 fat intake and reducing consumption of damaged omega-6 fats (think processed vegetable oils) to balance your omega-3 to omega-6 ratio, is also vital for your brain's overall health. Things such as krill oil, fish oil (krill oil also contains astaxanthin, which gives it an edge over fish oil for some people), coconut oil, and other healthy fats are very important for your brain's functions.

Exercise.

Another thing most of us are already tired of hearing, physical exercise is one of the best mental exercises. Look at it this way, if our brains are computers and learning things is our way of "uploading software" to them, then physical exercise is how we "upgrade our hardware." If you are having memory problems, if you are worried about your IQ, or if you generally believe that you are capable of more – start exercising. The effects on your body will be visible, and the effects on your mind will be drastically noticeable. So, after you're done practicing some of the methods and tips we outlined above, get up and do 30 minutes of exercise – your brain will be grateful.

Get a good night sleep every night.

Sleep is also notoriously crucial for our memory. We really shouldn't need to explain why either – when you haven't slept well you can barely think. When you've had a good night sleep, your brain is rested and ready for action. And as far as coffee goes – it's one of those things like sugar that has a positive short-term effect, but an awful long-term effect on our mental faculties.

Stop multitasking.

Multitasking is a great skill to have, but it should be used in moderation. We are so forced to multitask nowadays – at work, at school, on social media – that a lot of us are losing our abilities to focus. You can't have a well-functioning and effective memory if you can't keep your focus on one thing for at least a while. Start improving your focus and your mind will quickly follow.

Master new skills.

Speaking of keeping your focus, try mastering new skills. This is the best way to train your memory as it requires you to memorize further information. The more mentally-stimulating the skills are, the better. Mental math can be one such thing, but you don't need to constrain yourself to just that – there are a lot of cool skills you can toy with.

Schedule your work and studies properly.

One crucial mistake that a lot of us make both in school and at work is cramming all our work at the end of our deadlines. This is when we are usually most productive because we are under pressure, but it is also very unhelpful for our memory prowess. Learning and memorizing new things when we are under pressure us highly ineffective. Instead, when we properly schedule our work and studies, we utilize and improve our memory much more.

Use your memory in every situation.

The simplest advice is always to use your memory whenever you find a situation for it. Do you have to type a phone number that you just heard by ear? Try to memorize and sort it at once, instead of on batches of 2 or 3 digits. Do you need to type a different 8-digit, 10-digit or more extended code? Try to do it all at once. It's ok if you miss it a couple of times – that's how we learn and train. Try and remember long shopping lists without writing them, try and remember names and phrases by associating them with something, instead of writing them down or forgetting them.

Methods of becoming better at mental math as quickly and efficiently as possible

We learned quite a few different ways to solve mental subtraction multiplication and division equations. Some quicker, but more complicated, others slower but safer and better for training. However, how does one go about mastering them, learning how to solve complex problems with ease, and overall improving one's mental prowess? It's easy to say that exercise, and healthy eating is beneficial and vital for your memory and thought processes, but what are the active things and practice methods you can do to speed up the whole ordeal?

Practice, practice, practice.

As they thought us in school, the first and most important thing is to sit down and practice. There are lots of tips and tricks that make solving mental math problems more manageable, but even with them, you still need to put in the work. We'll try to avoid clichés like "Nothing good is easy" and other similar ones, but the core idea behind them is right – if you are set on becoming a human calculator and improving the way your brain works, get ready for some heavy lifting.

Read more math books.

We've tried to make this book as concise and useful for you as possible, but you can't expect to become a math expert with just one book. Find some more great math books and textbooks and start going over them. Repetition is the mother of memory, and that's especially true in math. Even just an old math textbook with some practice math equations can be of great use to you. Plus, once you master the mental math tips and methods we went through here, you'll likely want to learn even more complex problem-solving math skills. It may not feel like it now, but once you get the hang of it, math can be very addictive.

Tutor on the side.

One of the best methods to learn is to be taught. You may be quite good at learning through reading and not need a tutor, but the chances are that a good math teacher can significantly speed-up your learning process. Especially when it comes to solving practice equations, a tutor by your side can be a great benefit in keeping your focus and on track. Plus, once you're ready to move on beyond things like mental multiplication and division, the value of a tutor by your side will only increase as things become more complicated.

Play mind games.

Math doesn't need to be taught just through numbers. There are a plethora of excellent mind games you can practice with – both alone and with friends. As a rule of thumb, we advise against paying for some of the more overhyped mental browser games out there. Not so much because they don't work – any game that requires you to push your thoughts and memory beyond your everyday practices is a good thing. But there's usually no point in paying for a simple browser mind game when there are a ton of much more useful and much more fun mind games you can play – both offline and online.

Tricks & Tips

We went through quite a bit of information here, so why not cap things off with a quick list of some more tip and tricks we may have missed. In reality, there are a ton of other cool tips and tricks out there. Math is an incredibly extensive subject that extends over all exact sciences and has been perfected for millennia. As such, the amounts of methods you can do in math are virtually incalculable, as ironic as it may seem. Here, however, we mostly talked about mental subtraction, multiplication, and division, which are the basis of the arithmetic field and just a starting point of getting into more complex math problems. As such, here are some of the more basic tips and tricks you might want to master first:

Calculate from left to right, not the other way around.

Yes, we've spent years in school having the whole "calculate from right to left!" thing drilled in our heads over and over again. And hey – it's true. When you're calculating with a pen and paper, you should always do it from right left. It's clear, simple, and ensures that you won't make a mistake or forget something.

However, when it comes to mental math, adding and subtracting from right to left is impractical. It's slow, it's clumsy, and it requires you to carry numbers regularly. In contrast, calculating from left to right is much more intuitive. Yes, it's less risk-free on paper, but we are also not students anymore, are we? Well, some of you may be, but that's not the point. What's important is that with just a little bit of practice, you'll quickly learn to calculate mentally from left to right without making any mistakes. What's even cooler is that when you become better at it, you'll see how you'll start guessing even big numbers after the first one or two digits to the let – that's how intuitive mental math can become when done right and practiced enough.

Simplify whenever possible.

Math may be hard, but that's just even more reason to try and make it simpler whenever you can. All good things in life are simple, and we'd include simplified math equations in that! There are many different ways for you to "simplify" one math equation – to split some parts of it into their components, to divided others into their factors, to divide all components on a common denominator, to multiply them to a point where they become simpler to solve, etc., etc.

As we saw, even rounding numbers up or down is a perfectly acceptable strategy at times – as long as you keep track of the changes you're making and you're not doing any wrong actions in your calculations, you should be perfectly fine.

What's most beautiful about learning to simple math equations is that once you get used to doing it, even more, complex math problems will start seeming simple. All those nightmarish equations in math class that used to terrify us will soon begin to look exceedingly simpler when you make it a habit of simplifying every math problem you face. Pretty much the first thing to go through your mind should always be "Ok, how do I make this simpler?"

Memorize building blocks.

"Memorizing" is a word that a lot of people don't like, but we'll be using it in the next several points. And your frustration is completely understandable – after all, why learn tips and tricks if you're still going to have to memorize stuff, right? Well, with complex math problems, remembering some things is the best trick you can master.

When we talked about mental multiplication and division, we pointed out that before you start, you should make sure that you've memorized the primary multiplication table. And that's entirely true – knowing the multiplications of all basic numbers (up to 9 x 9 = 81) should be something you can do in your sleep. The multiplication table is not the only necessary thing you should memorize, however.

The mathematician and founder of ClearerThinking.org, Spencer Greenberg, is famous for saying that "by memorizing the basic "building blocks" of math, we can instantly get answers to simple problems that are embedded within more difficult ones." So, if you've forgotten about fractions, decimals, 1/n tables (like 1/6 is 0.166, 1/3 is 0.333, etc.), and percentages from school, you should remind yourself of them. Here's a quick reminder:

| Fraction | Decimal | Percent |
| --- | --- | --- |
| 1/2 | 0.5 | 50% |
| 1/3 | 0.333... | 33.333...% |
| 2/3 | 0.666... | 66.666...% |
| 1/4 | 0.25 | 25% |
| 3/4 | 0.75 | 75% |
| 1/5 | 0.2 | 20% |
| 2/5 | 0.4 | 40% |
| 3/5 | 0.6 | 60% |

4/5   0.8   80%

1/6   0.1666...   16.666...%

5/6   0.8333...   83.333...%


Memorize all the different ways you can solve an equation.

We went through quite a bit of different methods and ways to solve mental math equations. Properly mastering and memorizing every one of them is essential if you are ever to become a human calculator. Each mathematical problem has different solutions, with some being better for that particular problem and others – worse for it, but better for another equation.

It's not uncommon for people to start fancying one method for solving mathematical equations and to ignore all other ways. And while this can feel great at first, because you're solving everything that's laid in front of you with just one simple mechanic, in reality, your skills with all other methods and math solutions are rusting. And once you are faced with a different problem, you may run into difficulties.

Don't be satisfied with the several methods we listed here either – there are lots of other neat tricks and solutions to popular math problems. The more you practice and read, the more exciting tricks you'll find out.

For example, you'll quickly notice that multiplying by 12 is pretty much always easy – all you need to do is multiply your multiplicand by 10 and then add the multiplicand two times to the result. It looks like this:

4 x 12 =

4 x 10 + 4 + 4 =

40 + 8 = 48

It's a very simple trick, so we didn't bother including it above, but it works. Besides, as Mr. Greenberg says "we can instantly get answers to simple problems that are embedded within more difficult ones," meaning that by getting used to simple mechanics like this, you can apply them to components of much more significant math problems.

Another similar trick can be found when multiplying by 16. First, multiply the multiplicand by 10, then multiply it again by 5 (or multiply half of the multiplicand by 10), then add the two results together and add the multiplicand itself to them. Here's how it looks like in practice:

36 x 16 =

36 x 10 + 36 x 5 + 36 =

360 + 18 x 10 + 36 =

360 + 180 + 36 = 576

Memorizing as much of these tricks will make even the most complex equation much simpler, as it is made up of smaller components.

Memorize the basic squares.

As with the "building blocks" we mentioned above, memorizing the basic squares of numbers is a huge benefit for your mental math skills. Squares are the results of numbers when multiplied by themselves, and they are one of those things that can look terrifying when you first encounter them, but once you understand them, they become much more straightforward. And once you memorize them – a ton of different math problems become exceptionally easy.

The thing about mental math is that the larger the numbers become, the more complex the operations that solve them become as well. So, no matter how well you master the methods we mentioned above, sooner or later you'll reach a point where you'll need to use squares.

"Take the two numbers you're multiplying and think of them as their average, x, plus and minus the difference between each and their average, ±y," a mathematician from Askamathematician.com says. "These two numbers are squared, so rather than memorizing entire multiplication tables you only memorize squares."

If you're confused, here a table of the basic square numbers for you to start with:

| Number | Squared |
| --- | --- |
| 1 | 1 |
| 2 | 4 |
| 3 | 9 |
| 4 | 16 |
| 5 | 25 |
| 6 | 36 |
| 7 | 49 |
| 8 | 64 |
| 9 | 81 |
| 10 | 100 |
| 11 | 121 |
| 12 | 144 |
| 13 | 169 |
| 14 | 196 |

15   225

16   256

17   289

18   324

19   361

20   400

Yes, the latter ten can feel daunting, but they aren't – it's just ten numbers on top of the basic multiplication table. As soon as you memorize those and start using them in your mental equations, you'll quickly start learning even bigger squares like:

25 squared = 625

36 squared = 1296

54 squared = 2916

And so on.

Once you've got those down, you'll start seeing what the mathematician from Askamathematician.com meant. Simply put, here's what he meant – imagine that you need to calculate 50 x 8. Now imagine that you don't know it's 400. Here's how you can solve this using squares:

First, figure out what is the average number between 50 and 8 – it's 29. We know that, because 50 – 8 = 42 and 42 / 2 = 21. And 8 + 21 = 29, just as 29 + 21 = 50.

Next, we determine the square of 29 by multiplying 29 x 29 = 841. From here, we need to square the difference from 29 to 8 and to 50, which is 21. The square of 21 is 21 x 21 = 441.

The last step is to subtract the square of 21 from the square of 29 and to get:

841 – 441 = 400, which is the answer of 50 x 8

Now, we can calculate 50 x 8 pretty easily without all this hassle, but there are other numbers which can be much harder to multiply without squares. It's situations like these where it's hugely beneficial to know how to work with squares.

## Chapter 7: Vertically And Crosswise

The vertically and crosswise method demonstrates another way to multiply numbers together. It is best used when you are unable to determine what bases to use when multiplying.

Let's take a look at how this works

23 X 87

Lay the sum out like so

23

87

The first thing we need to do is multiply vertically

3 and 7

23

|

87

So 3 X 7 = 21

21

Two is carried   Carry over and add to the next number

Now we need to do the crosswise part, which is 7 X 2 plus 8 X 3

23


87

________

40 21

And finally we do the vertically part again, to 8 and 2 8 X 2 = 16

23

|

87

So 2 X 8 = 16

204 02 1

23 X 87 = 2001

This method also stands true when multiplying by three numbers. Let's take a look

466 X 734

Lay the sum out like so

466

734

The first thing we need to do is multiply vertically 6 and 4

4   6   6

            |

7   3   4

So 6 X 4 = 24

        24

Now we need to go the crosswise part, which is 6 X 4 plus 6 X 3

4   6   6

7   3   4

____________

      44 24

The next step is to continue on with the crosswise part and extend it to all the numbers, like so

4       6       6


7       3       4


        44 24

The sum now becomes

4 X 4 plus 6 X 3 plus 6 X 7

16 plus 18 plus 42 = 76 and then add the carried over four from the sum above giving 80

4       6       6


7       3       4


        80 44 24

Again we continue on with the crosswise, this time we use 4 X 3 plus 6 X 7

4  6  6


7  3  4

6280 44 24

Finally our last step is to do the vertically part, 4 X 7

4   6   6

  |

7   3   4

So 4 X 7 = 28

34 62 80 44 24

466 X 734 = 342044

MULTIPLICATION SHORTCUTS

By now you should be finding these sums easy. That's because the method you have just learnt makes them easy. Well you are about to learn another way of making certain multiplications even easier!

This next method works when the numbers being multiplied have the same difference between then from the base reference number.

For example take 17 X 23

The base to be used here would be 20.

There is a difference of 3 between 17 and twenty and also another difference of three between 23 and 20.

So for this next method to work, the differences must be the same.

So how do we do the sum when using this method?

Simple. We take the base; in this case it's 20. We square it, 400 and then subtract the square of the difference between the numbers; in this case 3 squared which is 9.

400 – 9 = 391

But how do we square 20? Just square the 2 which is 2 X 2 and add two zeros on the end!

How about 146 X 154?

Well they both have a difference of four from the base of 150 so we would square 150 and subtract the square of four.

To square 150 just square 15 which is 15 X 15 and add two zeros on the end, giving 22500

Next you subtract the difference of 4 squared, which is 16. 22500 – 16, to do this take away 20 and add four giving 22484.


Multiplying numbers that add to ten

This method allows the person to multiply any two numbers that have the same base and the last digits add to ten.

For example let's look at 24 X 26

You can see that both numbers have the same base, in this case it is 20. The last two numbers add up to ten, 4 plus 6 = 10.

Well then you will be happy to know that there is an even quicker method of multiplying sums like these and you don't need to use the base number for it. This is how it works

24 X 26

We take the first number 2 and multiply it by one more than two.

One more than 2 is three so it's 2 X 3 = 6. Finally we multiply the 4 (from the 24), and the 6 (from the 26), together giving 24. Bring the numbers together

24 X 26 = 624

Let's try 41 X 49

Like before, we multiply the base 4 by one more than it, 5 giving 4 X 5 = 20 Then we multiply the 1 (from the 41), and 9 (from the 49) together giving 9. Bring the numbers together

41 X 49 = 2009

Remember though that this 9 is in fact 09 as we need two digits after the 20. Think about it, when using this method whether the two numbers add up to ten, the only number when multiplied together that is less than ten is 1 X 9

If we were to do 2 X 8 (remember they must add to ten and have the same base) we get sixteen. So if the sum was 42 X 48 the answer would be 2016!

42 X 48

Multiply the 4 by one more than it, 5 giving 4 X 5 = 20

2 X 8 = 16

Bring the numbers together 42 X 48 = 2016

You shouldn't be surprised to learn that this method will also work with three digits

108 X 102

Take the 10 in 108 and multiply it by one more than it 11 giving 10 X 11 = 110

Multiply the 8 and 2

8 X 2 = 16

Bring the numbers Together 108 X 102 = 11016

Try out a few out for yourself and be amazed at how simple they are!

# MULTIPLYING BY 11 SHORTCUT

This neat little trick will show you how to multiply any two numbers by 11 in seconds!

Let's take, for example, 63 X 11

You take the number being multiplied and you split the numbers apart, like so

6   3

The next step is to add the two numbers together and put that answer in between the two numbers,

6 + 3 = 9

6 9 3

So 63 X 11 = 693

How easy was that?!

There is a slight variation when doing this trick when the numbers you are adding together are over 10.

Lets try 29 X 11

Again we separate the two numbers like so

2   9

Next we add the two together 2 + 9 = 11

But we do not put 11 in between the 2 and 9 giving 2119 as this is wrong. We carry over the tens unit of 11 to the 2 and add them together giving 319 So 29 X 11 = 319

So how about numbers that have more than two digits in them, like multiplying 234 X 11?

Well, this can be done just as easily you will be happy to know!

This is how it is done.

Separate out the first and the last digits of the number you are multiplying by 11

2    4

We don't forget about the middle number, 3 as this is the number we add to both separated digits:

2 (2 + 3)    (3 + 4) 4

So what we do here is put down are initial separated number, in this case it is 2, then the sum of 2 + 3 which is 5, then the sum of 3 + 4 which is 7 and finally the last separated number which is four.

234 X 11 = 2574

Let's try 566 X 11

Separate out the first and last, just like in the last example:

5    6

Now we add the middle digit 5 to the first and the last.

5 (5 + 6)    (6 + 6) 6

Both of these numbers go over ten so just be careful that you add an extra number where it is required.

5 (5 + 6 and 6 + 6) 6

giving

5 (11, 12) 6

here we add the 1 in the 12 to 11, giving twelve.

5 (12, 2) 6

and then we add the 1 of the 12 to the 5 giving

6 (2, 2) 6

then just put all the numbers

together 566 X 11 = 6226

Let's do one more example. 193 X 11

Separate the first and last digits

1   3

add nine to the first and to the last

1 (1 + 9)    (9 + 3) 3

add the sums together

1 (10, 12) 3

move the 1 in the twelve and add it to the 10.

1 (11, 2) 3

take the first 1 in the 11 and add it to the first 1 outside the brackets

2 (1, 2) 3

Finally put all the numbers

together 193 X 11 = 2123

Multiplying by 11 questions

a) 12 X 11

b) 45 X 11

c) 342 X 11

d) 546 X 11

e) 837 X 11

f) 32 X 11

g) 88 X 11

h) 445 X 11

Answers:

a) 132

b) 495

c) 3762

d) 6006

e) 9207

f) 352

g) 968

h) 4895

MULTIPLYING BY 9's

9 times tables are looked upon as being difficult by most people and with their current method for

calculating the answer it doesn't seem surprising. You are about to learn a technique to simple you will be surprised you didn't know it already!

Multiplying a single number by 9

Let's look at 6 X 9

The way to find the answer for this sum is that same as multiplying any single digit by nine.

Bring the 6 down one, to make 5. How many is five from 9? 4. Put the two numbers together, 6 X 9 = 54

How about 7 X 9?

Again bring the seven down one, making 6. How many is 6 from nine? 3. Bring the two numbers together,

7 X 9 = 63


Multiplying two digit numbers by 9

This method is slightly different but definitely no more complicated. Let's try 24 X 9

We look at the first digit, 2. We then add a one to it, giving 3. We then subtract 3 from 24 giving 21

Now we look at the 4 of 24 and find the difference between that and ten. 6 is the difference.

Bring the 21 and the 6

together. 24 X 9 = 216

Another, 43 X 9

Bring up the 4 to 5, and subtract 5 from 43 giving 38. Look back at the 43 in the sum, what is the difference between the 3 and ten? 7. Bring the 38 and 7 together

43 X 9 = 387

73 X 9?

Bring the 7 up one, to 8. 73 − 8 = 65

Look back at 73, what's the difference between the 3 and ten? 7. Bring the 65 and 7 together

73 X 9 = 657

Multiplying by 9 questions

a) 42 X 9

b) 5 X 9

c) 9 X 9

d) 66 X 9

e) 46 X 9

f) 94 X 9

g) 33 X 9

h) 57 X 9

Answers:

a) 378

b) 45

c) 81

d) 594

e) 414

f) 846

g) 297

h) 513

Multiplying a single digit by 99

Multiplying by 99 is not a lot different from multiplying by nine. Let's start by looking at an example.

4 X 99

Bring the 4 down one, to 3.

As there is only one digit we are multiplying we keep the first 9 as it is. What is the difference between four and ten? 6

Put all the numbers together

4 X 99 = 396

Let's take a look at 8 X 99

Bring the 8 down one, to 7.

What is the difference between the eight and 10?

Two. We only have one digit to multiply 99 by so we keep the 9.

Bring all the numbers

together 8 X 99 = 792

Multiplying two digit numbers by 99
So how do we go about multiplying two numbers
by 99?
23 X 99
We bring the 3 of 23 down by one, to 22.
Remember the rule "all from 9 and the last from
ten"? Well that rule applies here.
We have deducted the one to give us 22. Apply
the all from nine rule to 23, giving 7 and 7.
Bring all the numbers
together 23 X 99 = 2277
42 X 99
Subtract one from 42 giving 41. Now apply the all
from nine... rule, 42 all from 9 is 5 and 8
Bring all the numbers
together 42 X 99 = 4158
How do we calculate sums with more than two
nines? Read on.

Multiplying a single digit by 999
You won't be surprised to learn that multiplying
by more than two 9's is as easy as 99.
This is how it's done

4 X 999

Again, as we did for single numbers, bring the number down by one, giving 3.

As we only have one digit we don't touch the first two 9's, 99.

For the last part we find the difference between 4 and 10.

Bring all the numbers together 4 X 999 = 3996

7 X 999

Subtract one from 7 giving 6.

Find the difference between 7 and 10 giving 3.

Bring them all together,

7 X 999 = 6993

Multiplying two digit numbers by 999

This method is very similar to multiplying by 99.

62 X 999

Subtract one from 62 giving 61.

We have two digits in our sum, 6 and 2. So we leave the first 9 alone and apply the all from 9 and last from 10 rule.

62 gives 3 and 8.

Bring all the numbers together 62 X 999 = 61938

48 X 999

Subtract 1 giving 47. Keep the first nine and apply the all from nine and last from ten rule to 48 giving 5 and 2.

Bring all the numbers together 48 X 999 = 47952

This same method can be applied to three digit numbers

467 X 999

Subtract 1 giving 466

As there are three digits there are no nines left over in the 999. All from 9 and last from ten rule is applied to 467 giving us 533

Bring all the numbers together

467 X 999 = 466533

How about

467 X 99999?

Subtract one from 467 giving us 466

There are three digits and 5 nines so we leave the first two nines alone. Apply the all from 9 and last from ten rule again to 467 giving 533.

Bring all the numbers together

467 X 99999 = 46699533

You can see how easy these are can't you? With a little bit of practice you can do these in your head and in no time at all!

5 SQUARED SHORTCUT

This method is brilliant for squaring any two digit number that ends in five.

Here goes

$35^2$

From the methods you have learnt before you could tackle this sum like so

(30) 35 X 35

But there is an even easier way than using a base of 30.

Take the 3 and $5^2$ and separate them

3    52

Add one to 3 and you get four. 4 X 3 = 12

 Bring the 12 and $5^2$ together ($5^2$ is 25).

12 and 25

Therefore

$35^2$ = 1225

$75^2$

Add one to 7 giving 8.

8 X 7 = 56

Bring 56 and $5^2$ together

$75^2 = 5625$

5 squared shortcut questions

a) $45^2$

b) $55^2$

c) $65^2$

d) $35^2$

e) $95^2$

Answers:

a) 2025

b) 3025

c) 4225

d) 1225

e) 9025

SQUARING TWO DIGIT NUMBERS

By now you shouldn't be surprised to learn that there is a shortcut for squaring two digit numbers.

Let's look at $26^2$

Below are three steps you must take to calculate the answer

Square the first

2 times the first and last

Square the last


Ok so let's work through each step

$26^2$

Square the first, so $2^2 = 4$

2 X first and last, (2 X 2) X 6 =24

Square the last, $6^2 = 36$

So let's put all the numbers together

 4, 24, 36

The last digit in the answer is 6, so add the 3 to the 4 in 24 giving 27. Add the two of 27 to the four giving 6.

You could lay the numbers out one above the other

4

24

 36

_______ Add the numbers

676

Another, $57^2$

Square the first, $5^2 = 25$

(2 X 5) X 7 = 70

$7^2 = 49$

Lay the numbers out one above the other

25

 70

  49

_______ Add the numbers

3249

Try some out yourself, it won't take you long to get the hang of them.


Another way to square two digit numbers

You may have enjoyed using the previous method for calculating the square of a two digit number, well there is another method, just as easy, if not easier.

Let's take the same example again, $26^2$

The method here is to add to the sum whatever number it is above its base. So in this case we add 6 to 26

6 + 26 = 32

The base we were using in this example was 20, so we need to multiply 32 by 20. How? Times 32 by two and add a zero.

32 X 2 = 64 and add a zero, 640

Last part we need to add to 640 the square of the number above the base, 20. In this case 6 is above our base 20 so we square 6 and add it to 640

$6^2 = 36$

$640 + 36 = 676$

$26^2 = 676$


Let's try $42^2$

Our base here is 40

Add two to 42 as it is 2 above the base giving 44

Now multiply 44 by 4 giving 176 and add a zero as we multiplied by 40 giving 1760

Now square the number we added to 42 which was 2. Two squared is 4

Bring the numbers together

$42^2 = 1764$

That way was easy wasn't it? But did you notice that $42^2$ was near the base 50? So we could have actually used 50 as our base. How? Like so

$42^2$ and our base is 50 so 42 is eight less than fifty so we subtract 8 from 42 giving 34. So how do we multiply 34 by our base 50? Just multiply it by 100 and divide it by 2

34 X 100 = 3400 and then divide by 2 diving 1700

What number did we subtract from 42? We subtracted 8 so we now need to square this number and add it to 1700

$8^2 = 64$

64 + 1700 = 1764

$42^2 = 1764$


DECIMAL MULTIPLICATION

Multiplying a number that has decimal points in it may seem quite difficult to do. But with the methods you have already learnt you will see how easy they are to do.

Let's take 1.4 X 23

So how would we approach this sum? First thing when multiplying with a decimal point in the sum is to just forget about the decimal all together. In this case, 1.4 would become 14. So what we did was just remove the decimal. 3.6 would become 36 and so on.

So can you see what we are going to do now? We look at the sum again and it now looks like this:

14 X 23

So like before we choose a base that is common to both. Ten and two.

(10 X 2)    14 X 23

Again get the numbers that differ from the base.

43

(10 X 2)    14 X 23

Just like previous examples we multiply the four by the base 2, giving 8. We put this eight above the four.

8

43

(10 X 2)14 X 23

Add the 8 to 23 giving 31. Multiply 31 by the base 10 giving 310.

Just like before multiply the 4 by 3 giving twelve and add this to 310.

12 + 310 = 322.

Ok so this would otherwise be our answer for 14 X 23 but remember the original sum was 1.4 X 23 so how do we figure out what to do with the decimal?

Easy. How many numbers are to the right of the decimal point? The answer is 1. So from the answer 322 place a decimal point one place from the right.

1.4 X 23 = 32.2

If we were multiplying 1.44 X 23 we would place the decimal point two places from the right as there are two numbers to the right of the decimal point in the original sum.

Let's try 2.4 X 3.6

So like the last example we just ignore the decimals completely, giving us 24 X 36 and calculate it like you have learnt.

Use the base 10 and three.

(10 X 3) 24 X 36

Find the difference from the bases:

146

(10 X 3) 24 X 36

Multiply the 14 by the base 3 giving 42

42

146

(10 X 3) 24 X 36

Add this 42 to 36 giving 78 and multiply by 10 our base.

78 X 10 = 780

Multiply 14 X 6 giving 84 and add this to 780 giving 864

24 X 36 = 864

Remember we were originally multiplying two numbers with decimals in them; 2.4 and 3.6

In total, how many digits are there to the right of each decimal point?

With 2.4 there is one and with 3.6 there is one also. Add these two together so there is a total of two decimal points between them.

So looking back at our answer of 864, place a decimal point two digits in from the right, giving us 8.64

So 2.4 X 3.6 = 8.64

Let's try one more example

1.08 X 3.87

So just like before ignore the decimal points for the time being so our sum now looks like this:

108 X 387

Find our bases

100 and three as 100 X 4 will give us 400 which will work out well for 387

(100 X 4)   108 X 387

Find the difference between the numbers and their bases

8

(100 X 4)108 X 387

13

Now we multiply the 8 by the base 4 giving us 32

32

8

(100 X 4)108 X 387

13

So now we add 32 to 387 giving us 419

Multiply 419 by the base 100 giving us 41900

Now we multiply 8 by 13 giving us 104 and subtract this from 41900

108 X 387 = 41796

Remember we subtract as the 13 we multiplied by 8 carries a negative value as it was 13 below the base 400.

So now we go back to the original sum which was

1.08 X 3.87

How many numbers are to the right of 1.08? Two. How many to the right of 3.87? Two also. Add these together giving 4. So now we know that we have to have four numbers to the right of the decimal point in our answer:

1.08 X 3.87 = 4.1796

What about 10.8 X 3.87? Well there is one number to the right of 10.8 and there are two to the right of 3.87. Add these together giving us three. So we need to have three numbers to the right of the decimal point in our answer 41796

10.8 X 3.87 = 41.796

TIPS AND TRICKS

So what can you do with your new found mathematical abilities? Hopefully you will have found the shortcuts useful and have been applying them to your multiplications.

This section will demonstrate a few tricks that may be useful to you in your every day life.

How to convert Celsius to Fahrenheit

The general formula used to do this sum is far too complicated for what it needs to be. This little trick is far easier to do.

Let's try 11ºC. Convert it to Fahrenheit.

First step is to double the Celsius.

Second step is to add thirty.

11 X 2 = 22

Add 30 = 52

11ºC = 52º F

What about 22º C?

Double it and add thirty.

22 X 2 = 44

Add 30 = 74

22º C = 74ºF

So can we work out Fahrenheit to Celsius?

Just do the reverse!

Let's try 78ºF

Subtract 30.

Divide it by two.

78 − 30 = 48

48 halved = 24

78ºF = 24ºC

That's a nice easy way to figure out the temperature conversion, isn't it? Bear in mind this method does not give an exact temperature conversion but it provides us with an adequate result for practical purposes.

Converting kilograms to pounds

This little tip demonstrates an easy method of converting kilograms to pounds. The multiplying by 11 shortcut is needed to make this conversion simple.

Let's try and convert 80 Kg to pounds.

Use this method if the Kg ends in a 0 or 5

Double it, giving 160

Divide this by ten. To do this just remove the last zero, giving 16.

Multiply 16 by 11. Use the shortcut for multiplying by eleven; add the 1 and 6 and put the answer in between.

16 X 11 = 176

80 Kg = 176 lbs

Numbers that ended in a five or zero worked well as we were dividing by ten.

The proper conversion method is 1 Kilogram = 2.2 pounds.

So to do this sum we multiply the Kilograms by 0.2 and then by 11.

Let's try 12 Kg.

12 X 0.2 = 2.4

Multiply 2.4 by 11. Add the two and four together and put this number to the left of the decimal point giving 26.4

12 Kg = 26.4 lbs

Let's try another, 79 Kg

79 X 0.2 = 15.8

Now treat 15.8 X 11 as 158 X 11

Remember back to page 35 how to multiply by 11.

1(1+5) and (5+8)8

1(6) and (13)8

The one in 13 carries over to the 6.

1(7) and (3)8

Bring them all together

158 X 11 = 1738

Remember that we were initially calculating 15.8 X 11 and how many numbers are to the right of the decimal point in 15.8? One. So place a decimal point one place from the right of the answer 1738

15.8 X 11 = 173.8


CHECKING YOUR ANSWERS

You will be delighted to learn that there is a technique you can use to see if your answer is the right one or not.

Let's take a look at how we do this.

We have just calculated 17 X 22 and got the answer of 374

How do we know it is right?

Add up all the digits of our answer like so:

3 + 7 + 4 = 14 then add these digits, 1 + 4 = 5

So we keep adding the numbers until we get a single digit. In this case it is five.

So now we look at our sum, 17 X 22

Add each number of each part of the sum like so

1 + 7 and 2 + 2

giving us

8 and 4

Multiply these two numbers together giving 32

Add these two digits together, 3 + 2 = 5

Compare the two single digits; if they are the same you know your answer is correct. In this case we are right as we have 5 and 5.

Let's try 23 X 28

We get the answer of 644

So let's add all the digits up

6 + 4 + 4 = 14 and 1 + 4 = 5

Now look at the sum

23 X 28

2 + 3 X 2 + 8

5 X 10 = 50, 5 + 0 = 5

We get the same answer so our answer of 644 is correct.

SUMMARY SO FAR

I hope you have really enjoyed working through this book so far and also received something of value from it. The techniques you have learnt can be used in every day calculations and hopefully by now you can see the benefit of these mathematical techniques as opposed to your "usual" or normal methods of multiplication.

The next part of this book is the key to dividing.

Dividing by 9

Dividing by 9 has been seen by many as difficult to say the least. When dividing by nine a lot of people would divide by ten and use that answer as a rough guide when they are working with smaller numbers. This is fine as an estimate, but you will never get the correct answer this way.

But let me show you how to do it simply and effectively. What better way to demonstrate this than with an example.

9 / 71 is 9 divided by 71

I am now going to rewrite this sum:

9) 7/1

So what have I done here? Using the Vedic System to lay this sum out I have put the divisor (9) to the left of the sum with a bracket after it. This DOES NOT mean question 9 which 9) generally means.

So what is the 7/1 called and why is it written that way? The 7/1 is called the dividend. This is the number the divisor.

I have split the 7 1 into tens and units, 7/1

9)   7/1

What you do is bring down the tens unit (in this case 7), and put it under the single unit (in this case the 1).

9)  7/1

/7

_______

So can you see yet what needs to be done? What happens now is bring down the 7 from the tens units under the line, and then add the left hand side consisting of the 7 and 1...

9)  7/1

/7

_______

7/8

So what does 7/8 mean? Well it means that 9 goes into 71 seven times with a remainder of 8. Remember, the / does not mean divide by, it is merely separating the tens and units.

Let's try another one.

9 divided by 53

Lay it out in the Vedic fashion...

9)  5/3

Write down five under the 3

9)  5/3

/5

_____
Bring the five down under the line from the 5/3 part and add the 3 and five together

9)   5/3
/5

     _____
5/8

Therefore 9 divides into 53 five times remainder 8.

What about this one...

9 into 81

Lay it out in the Vedic style...

9)   8/1

Write down 8 under the units

9)   8/1
/8

     _____

Bring down the 8 from the tens and add the 8 and 1 together

9)   8/1
/8

     _____
8/9

So 9 goes into 81 eight times remainder 9. But is remainder 9 actually the correct   answer? Well if we have a remainder of nine then we have a spare nine left over, as 9 divides into 9 once remainder 0. So in the case when we have a remainder the same as the divisor we add one to the left hand number, in this case add one to the 8 giving 9 and a remainder of zero...

9)  8/1
/8

‾‾‾‾
9/0

So nine goes into 81 nine times remainder nothing.

Let's try 9 into 99

9)  9/9

Write down the 9 from the tens under the units

9)  9/9
/9

‾‾‾‾
Bring down the first 9 under the line and add the two 9's on the right side

9)  9/9
/9

‾‾‾‾‾‾
9/18

Nine goes into 18 twice with a zero remainder, so add 2 to the 9 giving 11

9)   9/9

/9

______

11/0

Nine goes into 99 eleven times no remainder.

You may have noticed how the remainder part can be determined by adding the two numbers from the dividend together.

The example of 9 goes into 72, add the 7 and 2 giving nine…

Or the 81 divided by 9, 8 + 1 = 9

Or 9 into 53, 5 + 3 = 8 (remainder 8)

So let's now try dividing 9 into numbers made up of more than two digits.

Take a look at 104 divided by 9.

Lay it out in the Vedic style as before:

9)   10/4

Now we don't bring down the 10 and put it under the four. We bring down the 1 and put it under the 0. So for all these examples we merely take the first number and place it under the second number:

9)   10/4

1/

What now? Well just like dividing nine by two numbers where we added the numbers on the right hand side, we do the same here, but in this case we add the 0 and 1 together and place the answer under the 4 on the right hand side.

9) 10/4

1/1

———

Next we just bring all the numbers down under the line, remembering to add as we go:

9) 10/4

1/1

———

11/5

What do I mean by add as we go? Look above. I brought down the first 1 and there was nothing under it so I didn't add anything to it. The next number, 0, I added to the 1 below it, and this gave me the 1 under the line. Finally I added the 4 and the one together giving me the five under the line.

Let's try another.

145 divided by 9.

9) 14/5

Bring down the one and put it under the four:

9)   14/5

1/

______

Add the 4 and one together and place under the 5

9)   14/5

1/5

______

Add the numbers in each column and place under the line,

9)   14/5

1/5

______

15/10

So now we have a remainder of ten. 9 will go into ten once with one remainder. Leave the remainder to the right and bring the 1 over to the left hand side, adding it to 15.

9)   14/5

1/5

______

16/1

Nine divides into 145 sixteen times with 1 remainder.

Let's try one more

Nine into 234

9)   23/4

2/

______

9)   23/4

2/5

______

Add and bring down under the line...

9)   23/4

2/5

______

25/9    (nine into 9 with no remainder...)

9)   23/4

2/5

______

26/0

9 goes into 234 26 times with no remainder.

What about dividing by numbers less than 9?

Dividing by 7

Let's look at dividing 7 into 46. We are going to use the same method as above except there is one slight adjustment to the calculation:

7)   4/6

When dividing by nine we notice that 9 is one less than ten. In this example 7 is three less than 10. We didn't need to include this part when dividing by nine as anything multiplied by 1 is the same.

7)  4/6

3 (this is the difference between 7 and ten).

7)  4/6

3

Next we multiply the 4 by the 3 and place it to the right hand side

7)  4/6

3    /12

7)  4/6

3    /12

_________

Add the numbers and bring below the line:

7)  4/6

3    /12

_________

4/18   (seven goes into 18 twice remainder 4)

7)  4/6

3    /12

_________

6/4

Seven goes into 46 six times remainder 4.

Dividing by 8

How about 8 into 134

8)   13/4

2     (remember, 8 is two from 10)

Multiply the fist number of the dividend which is 1 by the 2 and place it under the next number, in this case 3.

8)   13/4

2     2/

Add the three and 2 and place under the 4 as we have done before:

8)   13/4

2     2/5

_______

Add the numbers and bring under the line

8)   13/4

2     2/5

_______

15/9   (eight goes into 9 once, remainder 1)

8)   13/4

2     2/5

_______

16/1

Eight divides into 134 16 times with one remainder.

Simple Division

Division is deemed by many as more effort and harder to complete than multiplication.

Well let's eliminate that myth now with this instalment for division.

So, quickly let's do short division so as you can divide 6 into 76.

Layout the sum like so:

We need to ask ourselves what we need to multiply 6 by to give us 7? Well we can't multiply any whole number to give us seven, so we go for a number below 7. We can multiply 6 by 1 to give six and the difference between six and 7 is one (our remainder).

So next we need to ask what we need to multiply six by to give us 16. Nothing will give us sixteen directly, so we try 3. 6 X 3 = 18. Too big.

6 X 2 = 12. This works.

So we place the two in the part of our answer. What's the remainder? Take twelve from sixteen giving us a remainder of 4.

Let's try dividing 223 by 53
Lay out the sum like so:

You will notice the 3 of 53 is slightly smaller. This does not mean 5 cubed, it is merely written this way so as you can see it is separate from the 5. We can call this our FLAGGED value.
So now we need to ask ourselves how many times does 5 (from 53) go into 22 ( this first part of the sum, separated by those black lines).
5 goes into 22 four times with a remainder of two.
Note how the two remainder goes under the three.

Next we take that four and we multiply it by our flagged 3 that belongs to 53, giving us twelve.
So what next? Well we have our twelve which we just calculated, and we also have 23 from the remainder 2 and the three (above).
So we need to subtract 12 from 23 giving us 11.

Our answer now becomes:

4 remainder 11.
Pretty simple eh?!

Fractions

Whilst in school we all probably hated the thought of working with fractions. They seemed difficult, hard to grasp, and the techniques we learnt to calculate them was by no means easy.

You can rest assured the methods you are going to learn now make adding fractions a cinch.

Let's begin.

Adding Fractions

$1/4 + 1/3$

The method we use for adding ANY fraction is like so:

We cross multiply the fractions.

So what we have is 1 X 3 and 1 X 4

We add the results of both fractions together, so it will look like this:

(1 X 3) + (1 X 4) = 7

Next we need to find the denominator ( the bottom part of the fraction). This is easily done by multiplying 4 X 3, giving 12.

1/4 + 1/3 = 7/12

Ok let's try 2/5 + 4/7

Remember same rule again we cross multiply

As before, (2 X 7) + (4 X 5) = 34

Now we need to find the denominator, to do this multiply the two bottom numbers together, 5 X 7 = 35

2/5 + 4/7 = 34/35

What about this one:

2/5 + 5/7

(2 X 7) + (5 X 5) = 39

Bottom denominator 5 X 7 = 35

2/5 + 4/7 = 39/35

But this isn't necessarily the fraction in its lowest form. We can see that the denominator is less than the numerator (39) and that it will go into 39 once with a remainder of four.

So how do we work it out this way? Well as it will go into 39 once we put a one at the start of the fraction and make the remainder and numerator into the fraction.

14/35

2/5 + 4/7 = 39/35 = 14/35

Subtracting Fractions

Subtracting fractions is just as simple as adding them.

Let's take the same example as one above and walk through it:

1/4 - 1/3

Again we cross multiply but instead of adding we subtract

$(1 \times 3) - (1 \times 4) = 3 - 4 = -1$

Multiply the denominators together, $(4 \times 3) = 12$

This gives us $- 1/12$

The whole fraction becomes minus 1 over twelve. Make sure not to make the mistake of saying the 1 is negative, the WHOLE fraction has become negative.

Multiplying Fractions

We will continue on from using previous examples, but instead of adding or subtracting we will multiply.

Ok let's try 2/5 X 4/7

There is a slight variation to multiplying fractions. Instead of cross multiplying, we just directly multiply, left side with right side. In the case above, we work it like so:

(2 X 4) / (5 X 7) = 8 / 35

Let's try another:

1/4 - 1/3

Remember, just like before, multiply top by top, bottom by bottom:

(1 X 1) / (4 X 3) = 1 / 12

Dividing Fractions

Welcome to the last week of our adventure through Vedic Maths.

By now you are probably thinking why you were not shown this method in school. Good question.

But anyway, on with the final instalment.

How to divide fractions.

Well if you can remember what we did for multiplying fractions, which entailed multiplying the numbers directly across from one another. Well you will be glad to hear that the same rule applies for dividing fractions, but with one small adjustment.

Ok let's try 2/5 ÷ 4/7

Now, this sum can be laid out just like before when we were multiplying:

But what we do is swap the numbers around on one of the fraction and do as before for multiplying, so in this case it would be:

Then we multiply across:

2 X 7 for the top

5 X 4 for the bottom giving us 14/20 as our answer, which if halved gives 7/10 (lowest this will go).

Square roots

The symbol for square roots is this:√

Imagine multiplying 25 X 25 and you get the answer of 625. Well the square root is the reverse.

$\sqrt{625} = 25$

So how do we go about performing this reverse function?

Well it's not that difficult at all!

Let's try √2209

We need to separate the numbers, giving us 22 and 09.

Look at the 22. We know that the first number MUST be 4 as 4 X 4 = 16.

5 X 5 = 25 and this will be above our value 22.

So what's our remainder? 4 X 4 = 16. Sixteen from 22 is 6, our remainder.

√22 6 09 = 4?

To get the last part of the answer we divide 60 into TWICE the first part of the answer, in this case we use 4.

60 into twice 4 is 8 into 60 = 7.5

We can now drop the .5 from 7.5 as this can easily go back into the sum and cancel anyway, so we get the answer of 47.

√1089

Ok, we can see that the first part of the answer will be 3 as 3 X 3 = 9 which is below 10 by splitting 10 89 apart.

√1089= 3? There is a remainder of 1.

√10189= 3?

Twice 3 (6) then divide this 6 into 18, giving 3.

This is all the work you need to do! √1089 = 33

√2809 = ?

First part, 28. 5 X 5 = 25 (the first part of our answer).

25 from 28 is 3.

√28 3 09 = 5?

Double 5 giving ten and divide this into 30, which goes in 3 times with a remainder of 0.

√28 3 09 = 5 3?

√2809 = 53

Estimating Square roots

√70 = ?

First let's try to guess the square root, from what we did before.

8 X 8 = 64.

9 X 9 = 81

So we know that the first part of our answer starts with 8.

Now we divide our 8 into the 70 giving us 8.75,

70 divided by 8 = 8.75

So now we split the difference between 8.75 and our 8 estimate, giving 0.75

Now split the difference of 0.75 giving us 0.375

Finally add this result to our estimate 8 giving 8.375

The answer will always be slightly higher than the correct one but for estimating purposes it is pretty accurate!

Let's go one more:

√29 = ?

Again take an estimate, 5 looks good (5 X 5 = 25).

Divide 25 into 29 and it goes once with remainder 4.

Five divides into 40 eight times. This gives us 5.8

Split the difference between 5 and 5.8 giving us 5.4

The correct answer is 5.385 but 5.4 is a pretty good estimate seeing as that's what we are aiming to do!

√31 25 = ?

Pair off the numbers starting from the right,

31   25

Each pair has one digit answer to it.

Let's estimate the first digit of 31, 5 looks good. (5 X 5 = 25).

There is one more digit to the answer so we add one zero to our answer, 50.

To divide by 50 we divide by ten giving 5.

3125 divided by 10 is 312.5

We now divide by 5 giving us 62.5

So now we again split the difference 62.5 – 50 = 12.5

12.5 divided by 2 = 6.25

Round off downwards and add this to your first estimate of 50, giving 56

√3125 = 56


Quick Sums

12 X 7

first thing is to always multiply the 1 of the twelve by the number we are multiplying by, in this case 7. So 1 X 7 = 7.

Multiply this 7 by 10 giving 70.

Now multiply the 7 by the 2 of twelve giving 14.

Add this to 70 giving 84. Therefore 7 X 12 = 84! How quick was that!

Let's try another:

17 X 12

Remember, multiply the 17 by the 1 in 12 and multiply by 10 (just add a zero to the end): 1 X 17 = 17, multiplied by 10 giving 170.

Multiply 17 by 2 giving 34.

Add 34 to 170 giving 204.

So 17 X 12 = 204

Let's go one more

24 X 12

Multiply 24 X 1 = 24. Multiply by 10 giving 240.

Multiply 24 by 2 = 48. Add to 240 giving us 288

24 X 12 = 288 Simple!!

# Chapter 8: Teach Your Kids Arithmetic - Subtraction Shortcuts

As students, we become comfortable with what we learn first. Of the four arithmetic operations of addition, subtraction, multiplication, and division, we learn to add first and for this reason are most comfortable with addition. If we apply the principle of thinking in terms of what we are most comfortable with, then subtraction need not be a difficult operation to master. Conse🞏uently, by applying addition principles to subtraction, we find our shortcut to mastery of this operation.

Addition is the first pillar of arithmetic magic. Once this operation is mastered, all others can be con🞏uered. The reason for this is inherent in the nature and interrelationship of the four arithmetic operations. You see addition and subtraction are inverse operations of one another. This interrelationship imputes a connection between these two procedures. What this means is that we can master one through the mastery of the other. Moreover, multiplication can be mastered through addition, since multiplication is repeated addition; similarly, division via multiplication or subtraction, since division is repeated subtraction as well as the inverse operation of multiplication. Thus why fret with mastering four distinct arithmetic operations when we can think of mastering one and then using shortcuts and derivative principles to master the others?

Such is the case with subtraction. Here I show you how to handle subtraction problems like 106 - 53. Rather than do a subtraction, you can think of "adding up" from 53 to 106. Another way of thinking about 106 - 53 is what is missing from 53 to make 106? This is the same principle that is taught to cashiers to make change, before, of course, the newer cash registers came out that do this for them. What cashiers would do is count up from the 53 to the 106. Thus we count from 53 to 60 to get 7. Then we count from 60 to 100 to get 40 more. So far we have counted 7 + 40 or 47 total. The final step is to count from 100 to 106 which is 6 more. As a result, we have 47 + 6 or 53 as our total. Thus 106 - 53 is 53. Let us take one more example to show how nicely this shortcut works. Take 96 - 49. Rather than fumble with this, it's simple if we add up from 49 to 96 as thus: 49 to 50 makes 1; 50 to 96 makes 46; and 1 + 46 is equal to 47. So 96 - 49 is 47.

If you apply this subtraction shortcut regularly, you will never have a problem with this operation again. This method works with more complicated examples as well as you can easily verify. Try to teach this method to your kids and have them practice with a few examples. In addition to the good mental workout they will get, their arithmetic skills will soar to new heights. And there's nothing better than seeing those A's on their report cards.

Vedic Mathematics Vs Abacus - What Will Suit for Kid?

Abacus is a calculating tool which first originated in the European countries. However, it was in China where Abacus became popular and was used for day to day calculations. Predominantly, used as a calculating tool, it has a frame consisting of wires which are attached to frame and beads which slide along these wires. Each bead represents one unit.

Abacus is mainly used to perform addition, subtraction, division, and multiplication. It is suggested that abacus learning at a very young age is useful in actuating the brains of the kids. When a child works on abacus, he/she will simultaneously use both his hands to move the beads. The right hand actuates the left hemisphere and the left hand actuates the right hemisphere, thereby helping in developing both sides of the brain in a balanced way. This promotes rapid and balanced development of the entire brain of the child. It is also suggested that Abacus math should be started at very early childhood, as young as age 4. Eventually the child retains the memory of bead positions and the relevant notation.

Abacus math if started during later ages can create a bit of hindrance.

• Although exceptionally helpful, abacus has plenty of drawbacks as the child might get overconfident in mathematics and the child might bypass the regular functions like addition, subtraction, multiplication and division methods.

• Abacus is primarily about cramming. It in a way creates monotony and takes well over two years to master it which might lead the child to get bored.

• Advanced mathematical concepts like calculus, algebra and geometry cannot be solved using abacus, an abacus in contrast to Vedic Mathematics is just basic and elementary.

Vedic Mathematics system is based on the 16 Vedic Sutras. These 16 Sutras were originally written in Sanskrit language and can be easily memorized and using these all kinds of calculations can be made. Vedic mathematics enables one to solve long mathematical problems quickly. It was founded in 1911 and has its roots in Atharva Veda. Vedic math can be entirely done in mind and paperwork is not required. Vedic math starts at a basic level of numbers and gradually progressing to simple additions, subtractions, multiplications and division.

Some advantages of using Vedic Math are -

• Vedic math is not just about solving the basic calculations as with Vedic math one can also be able to solve complex geometrical theorems, calculus sums and algebraic problems.

• Vedic math can be started at later ages as well without any difficulty.

• It is also very useful for competitive exams specifically while solving multiple choice questions where timing is an issue!

The rules of calculation are very simple; It focuses more on learning through logic and understanding of the fundamental concepts of mathematics rather than cramming and repetition as in the case of abacus. These formulae describe the way the mind naturally works and are therefore a great help in directing the student to the appropriate method of solution.

So, basically what a child does in Vedic mathematics is, he/she will derive the answers using the concepts of Vedic mathematics and then compare their final answers got by the regular mathematics process and that will help the child in understanding mathematics better.

One of the best aspects of learning and using Vedic mathematics is that it does not become an additional burden for students, teachers & parents. It rather complements the existing mathematics syllabus and makes mathematics more interesting and enjoyable for all. The only drawback of Vedic mathematics is that it is not advisable for kindergarten and primary school children and a child can understand its concepts only after a certain age; say after the age of 9 or 10. However the advantages and applications of Vedic mathematics are so wide that it minor drawbacks can be overlooked and should is preferred over abacus.

Mental Calculation Methods for Addition and Subtraction

This page shows some of the most common methods for mental addition and subtraction. It is not essential that students learn any individual method, but it is important that students learn a variety of methods from which they can choose to carry out mental calculations.

The notes below illustrate the basic calculations which would be carried out to use each method. It is not intended that students should learn to record these mental methods like this – a verbal report or actions on a hundreds chart or number line is more appropriate. There is no need to teach students the names of these methods.

Mental methods
Front-end estimation for addition and subtraction

Mental computation generally begins with the larger place value digits. This contrasts with standard written algorithms for addition and subtraction.

125 + 37 is approximately 12 tens + 3 tens so about 15 tens or 150.

1.25 + 0.37 is approximately 12 tenths + 3 tenths so about 15 tenths or 1.5.

125 − 37 is approximately 12 tens − 3 tens so about 9 tens or 90.

1.25 − 0.37 is approximately 12 tenths − 3 tenths so about 9 tenths or 0.9.

Better estimates could be obtained by more carefully rounding "to the nearest", but simple rounding up or down to a convenient number as illustrated here is often all that is required for practical purposes, including checking the reasonableness of calculator or other exact calculation.

Finding upper and lower bounds on the answer

Being able to find bounds between which an answer lies is a very useful skill.

256 + 136

Lower estimate: 200 + 100 = 300

Upper estimate: 300 + 200 = 500

So the exact answer must lie between 300 and 500.

2.56 + 1.36

Lower estimate: 2 + 1 = 3

Upper estimate: 3 + 2 = 5

So the exact answer must lie between 3 and 5.

456 − 136

Lower estimate: start with less and subtract more (400 − 200 = 200)

Upper estimate: start with more and subtract less (500 − 100 = 400)

So the exact answer must lie between 200 and 400.

4.56 − 1.36

Lower estimate: start with less and subtract more (4 − 2 = 2)

Upper estimate: start with more and subtract less (5 − 1 = 4)

So the exact answer must lie between 2 and 4.

Addition in stages and subtraction in stages

Students can split numbers to be added or subtracted into convenient components. What are convenient components depends on the numbers involved and on students' own number knowledge, but combining to the nearest multiple of ten or a hundred is usually advantageous.

To add 56 and 17, start with 56, and then add 4 to get 60. This leaves 13 to add, which is 10 plus 3. 60 + 10 is 70 and 70 + 3 is 73.
56 + 17 = 56 + 4 + 13 = 60 + 13 = 73

To add 5.6 and 1.7, start with 5.6, and then add 0.44 to get 6.0. This leaves 1.3 to add, which is 1.0 plus 0.3. 6.0 + 1.0 is 7.0 and 7.0 + 0.3 is 73.
5.6 + 1.7 = 5.6 + 0.4 + 1.3 = 6 + 1.3 = 7.3

To subtract 9 from 57, first subtract 7 (which gives 50) and then subtract 2, to get 48.
57 − 9 = 57 − 7 − 2 = 50 − 2 = 48

To subtract 17 from 205, first subtract 5 to get 200, and that leaves 12 to subtract. Subtract 10 from 200 to get 190, and then subtract 2 more to get the answer 188.

$205 - 17 = 205 - 5 - 12 = 200 - 12 = 200 - 10 - 2 = 190 - 2 = 188$

To add 1021 and 87, add 80 to 1020, which gives 1100. Then add the remaining 1 and 7 to get 1108.

$1021 + 87 = 1020 + 80 + 1 + 7 = 1100 + 8 = 1108.$

Adjusting the result of an easier calculation

- To add 9 to 65, first add 10, which is easy, and then subtract 1 to get answer 74.

- To add 0.9 to 6.5 , first add 1.0, which is easy,  and then subtract 0.1 to get answer 7.4

- To find 19 + 19, first find 20 + 20, which is 40, and then subtract 2 to get answer 38.

- To find 1.9 + 1.9, first find 2.0 + 2.0, which is 4.0, and then subtract 0.2 to get answer 3.8.

- To subtract 98 from 134, first subtract 100 to get 34, and then because too much has been subtracted, add 2 to get the answer of 36.

- To subtract 49 from 201, first subtract 50 (one too many) from 200 (one too few) to get 150 and then add the 1 from 201, and the extra 1 that was subtracted, so that the answer is 52.

## Chapter 9: Ten Times Tables

Learning up to 10 X 10 was a major stumbling block for many people in school. Sitting there reciting the tables hoping that repetition would eventually get the numbers to stick in the brain.

Maybe the numbers did sink in, maybe they didn't. When you learn this technique it won't matter whether they sank in or not!

This whole multiplication method is built around a "base". A base is a number that we can use as a reference point to make our multiplication easier. Let's try this base out.

For our first multiplication we will try 8 X 7

Using your old multiplication method you would probably recite 8, 16, 24, 32, 40, 48, and 56 and count these numbers on your fingers until you get to the 7th finger.

Using the new method you won't have to count on your fingers any more.

8 X 7

We need a common base between the numbers 8 and 7. What whole number are they close to?

10 is the closest to 7 and 8.

10 is called a "base" or reference number. If it is confusing you, call it what you like. Just remember that you need a number that is close to the numbers being multiplied. In this case it is 10.

If we were calculating 12 X 13 we can also use ten as it is close to them.


Ok, back to the sum 8 X 7

We use the base 10 so lay the sum out like this:

(10)8 X 7

How many numbers does it take to make the 8 into 10? Two.

How many numbers does it take to make the 7 into 10? Three

So put these numbers below the sum like so:

(10)8 X 7

  23

So why are we putting the numbers below the sum? This is because the numbers being calculated are below the base number, ten. As the numbers are below the base they carry a negative value.

So what's next? Now we have to subtract diagonally either 7 − 2 OR 8 − 3. It doesn't matter which one you choose to subtract, just make sure it is the easiest one. In this example subtracting 2 from 7 seems to be the easier choice.

So, 7 − 2 = 5

5 is the first part of the answer. Multiply this 5 by the base number of 10, giving 50.

So how do we get the second part of the answer? Let's look back at the sum again:

(10)8 X 7 = 5?

23

We now need to find the second part of the answer.

To do this we multiply the two numbers that make up 10. In this case it is the 2 and 3

2 X 3 = 6

6 is the second part of the answer. Put the two digits together and our sum is complete:

8 X 7 = 56

Let's try another. 6 X 12

(10)8 X 7 = 56

 23

Again we need to find a base number. 10 is close to both digits.

(10)6 X 12

Now find the number of digits it takes to make up ten.

2

(10)6 X 12

 4

So why is the 2 above the 12? The reason for this is because 12 is 2 above ten. So we put it above the 12. Six is 4 below ten so we put the four below the 6.

Next step we add or subtract diagonally. The reason why you may add is because the 2 is above ten giving it a positive value.

So you could add or subtract like so:

6 + 2 = 8 OR 12 − 4 = 8

So we have 8. We multiply this 8 by the base number 10, giving 80.

2

(10)6 X 12 = 8?

4

So how do we get the next number, seeing as one number is below and the other is above?

Simple. Multiply the 4 by 2. Note here that as the four is below, it has a − (minus) value.

So the sum is -4 X 2 = -8

What about before when the two numbers were below? Why were they not negative? The reason for this is because:

A minus times a minus = a plus
A minus times a plus = a minus
A plus times a plus = a plus

A good way to remember that minus times a minus is a plus is imagine a minus sign is a sword. Picture in your mind two swords clashing together. They look like a cross, or a plus sign on its side.
For minus times a plus is a minus imagine again the sword knocking the cross out of the way leaving just the one sword, a minus.
Ok so now we have to finish off the sum.

## Conclusion

So, that was fun, wasn't it? As you saw, mental math is only scary at first glance because of how much memory work it requires. In reality, however, once you've gotten the knack of the different methods of subtraction, multiplication, and division, and once you train your memory to operate on an adequate capacity – mental math is easy!

3-digit, 4-digit, 5-digit numbers – none of them need to pose a problem for you any longer. With those tricks up your sleeve, you'll be able to quickly deal with all kinds of mental issues without the use of your phone, a calculator, the internet, or even just a piece of paper. You'll be able to solve entire complex problems with just a glance, and you'll no longer have issues calculating interest rates, profits, percentages or taxes. By being so quick on your feet, you'll also start noticing good financial opportunities whenever they present themselves, and you'll become a much better professional, pretty much regardless of what your profession is.

What's most important, however, is that by learning to do mental math, you'll indirectly

drastically improve all of your mental processes. Memory, problem-solving skills, multitasking and keeping your focus, IQ, creativity, reaction time, and much more – all of those things will become easier and easier with enough practice.
Good luck!